Earthing and Bonding

GUIDANCE NOTE

On publication of the 17th Edition of the IEE Wiring Regulations (BS 7671), any amendments to this document will be freely available at www.theiet.org

IEE Wiring Regulations
BS 7671 : 2001 Requirements for Electrical Installations Including Amendments No 1 : 2002 and No 2 : 2004

Published by The Institution of Engineering and Technology, London, UK
© 2006 The Institution of Engineering and Technology
First Published 2006

The Institution of Engineering and Technology is the new institution formed by the joining together of two great institutions: the IEE (Institution of Electrical Engineers) and the IIE (The Institution of Incorporated Engineers). The new Institution is the inheritor of the IEE brand and all its products and services, such as this one, which we hope you find useful.

This publication is copyright under the Berne Convention 2003 and the Universal Copyright Convention. All rights reserved. Apart from any fair dealing for the purposes of research or private study, or criticism or review, as permitted under the Copyright, Designs and Patents Act, 1988, this publication may be reproduced, stored or transmitted, in any form or by any means, only with the prior permission in writing of the publishers, or in the case of reprographic reproduction in accordance with the terms of licences issued by the Copyright Licensing Agency. Enquiries concerning reproduction outside those terms should be sent to the publishers at the following address: The Institution of Engineering and Technology, Michael Faraday House, Six Hills Way, Stevenage, SG1 2AY, United Kingdom.

Copies may be obtained from:
The Institution of Engineering and Technology
P.O. Box 96
Stevenage
SG1 2SD, UK
Tel: +44 (0)1438 767328
Email: sales@theiet.org
www.theiet.org

While the author, publisher and contributors believe that the information and guidance given in this work is correct, all parties must rely upon their own skill and judgement when making use of it. Neither the author, the publisher nor any contributor assume any liability to anyone for any loss or damage caused by any error or omission in the work, whether such error or omission is the result of negligence or any other cause. Where reference is made to legislation it is not to be considered as legal advice. Any and all such liability is disclaimed.

ISBN 978-0-86341-616-3

Typeset in the UK by Wyndeham Pre-Press, London
Printed in the UK by Polestar Wheatons, Exeter, Devon

Contents

Cooperating organisations ix
Acknowledgements x
Preface xi

Introduction 1

Chapter 1 Protective earthing 3

1.1   Protective earthing 3
1.2   Source earthing 3
1.3   The purpose of source earthing 3
1.4   The purpose of electrical equipment earthing 4

Chapter 2 The means of earthing 5

2.1   The means of earthing 5
2.2   Distributors' facilities 5
2.3   Earth electrodes 6
2.4   Rod electrodes 6
2.5   Tape and wire electrodes 8
2.6   Plate electrodes 8
2.7   Structural metalwork electrodes 9
2.8   Metal covering of cables for electrodes 10
2.9   Location of the installation earth electrode 10
2.10 Resistance of the earth electrode 10
    2.10.1 Calculating the resistance of a rod electrode 12
    2.10.2 Calculating the resistance of a plate electrode 13
    2.10.3 Calculating the resistance of a tape electrode 13
    2.10.4 Limiting values of the electrode resistance for TT systems 13
2.11 Electrode installation 14
2.12 Electrode loading capacity 15
2.13 Measurement of the electrode resistance 15
    2.13.1 Proprietary earth electrode test instrument 15
    2.13.2 Earth fault loop impedance test instrument for measuring electrode resistance 17
2.14 Determination of the external earth fault loop impedance, Z_e 18
2.15 Responsibility for providing a means of earthing 19

Chapter 3 The earthing conductor 21

3.1   The earthing conductor 21
3.2   The cross-sectional area of an earthing conductor 22
3.3   The CSA of a buried earthing conductor 23
3.4   Minimum CSA of an earthing conductor 24

3.5	Impedance contribution of an earthing conductor	24
3.6	Colour identification of an earthing conductor	24
3.7	Protection of an earthing conductor against external influences	24
3.8	Disconnection of the earthing conductor	25
3.9	Connection of the earthing conductor to the means of earthing	25

Chapter 4 System types and earthing arrangements — 27

4.1	A system	27
4.2	TN-C system	28
4.3	TN-S system	29
4.4	TN-C-S system	30
4.5	TT system	32
4.6	IT system	34

Chapter 5 Main equipotential bonding — 35

5.1	The purpose of main equipotential bonding	35
5.2	Main equipotential bonding conductors	38
	5.2.1 Cross-sectional areas	41
	5.2.2 Identification	42
	5.2.3 Supports	43
	5.2.4 Alterations and extensions	44
5.3	Earth-free equipotential bonding	45
5.4	Bonding of lightning protection systems	46
5.5	Extraneous-conductive-parts common to a number of buildings	47
5.6	Installations serving more than one building	49
5.7	Multi-occupancy premises	52

Chapter 6 Extraneous-conductive-parts and their connections — 55

6.1	Definition of an extraneous-conductive-part	55
6.2	Some examples of extraneous-conductive-parts	57
6.3	An example of a conductive part which is not an extraneous-conductive-part	57
6.4	Connection to pipework	58
6.5	Connections to structural steel and buried steel grids	61

Chapter 7 Automatic disconnection — 63

7.1	Automatic disconnection	63
7.2	TN systems	63
	7.2.1 Earth fault loop impedance	65
	7.2.2 Mixed disconnection times	69
	7.2.3 Automatic disconnection for socket-outlets etc. in 5 s	72
	7.2.4 Automatic disconnection using an RCD	73
7.3	TT systems	73
7.4	IT systems	73
7.5	Automatic disconnection for portable equipment for use outdoors	75
7.6	Automatic disconnection for circuits supplying fixed equipment outdoors	75
7.7	RCDs in series	76

7.8	Automatic disconnection for reduced low voltage circuits	77
7.9	Automatic disconnection and alternative supplies	79
7.10	Separated extra-low voltage systems (SELV)	80
7.11	Protective extra-low voltage systems (PELV)	80
7.12	Functional extra-low voltage systems (FELV)	81

Chapter 8 Supplementary equipotential bonding — 83

8.1	Supplementary equipotential bonding	83
8.2	Supplementary bonding conductor types	85
8.3	CSAs of supplementary bonding conductors	86
8.4	Limitations on resistance of supplementary bonding conductors	87
8.5	Supports for supplementary bonding conductors	87
8.6	Bath and shower rooms	88
8.7	Shower cabinet located in a bedroom	91
8.8	Swimming pools	91
8.9	Agricultural and horticultural premises	94
8.10	Restrictive conductive locations	97
8.11	Static inverters	99
8.12	Other locations of increased risk	99
8.13	Where automatic disconnection is not achievable	99

Chapter 9 Circuit protective conductors — 101

9.1	Circuit protective conductors	101
9.2	Cross-sectional areas	102
	9.2.1 Calculation of CSA – general case	102
	9.2.2 Selection of CSA – general case	103
	9.2.3 Non-copper CPCs	104
	9.2.4 Evaluation of k	105
9.3	Armouring	106
	9.3.1 Calculation of CSA – armoured cable	106
	9.3.2 Selection of CSA – armoured cable	106
	9.3.3 Contribution to earth fault loop impedance	107
	9.3.4 Amouring inadequate for a CPC	107
	9.3.5 Termination of armoured cables	107
9.4	Steel conduit	108
	9.4.1 Calculation of CSA – steel conduit	109
	9.4.2 Selection of CSA – steel conduit	110
	9.4.3 Maintaining protective conductor continuity	111
9.5	Steel trunking and ducting	111
	9.5.1 Maintaining protective conductor continuity	112
9.6	Metal enclosures	112
9.7	Terminations in accessories	113
9.8	CPC for protective and functional purposes	115
9.9	Significant protective conductor currents	115
	9.9.1 Equipment	118
	9.9.2 Labelling at distribution boards	118
	9.9.3 Ring final circuits	119
	9.9.4 Radial final circuits	119
	9.9.5 Busbar systems	120
	9.9.6 Connection of an item of equipment (protective conductor current exceeding 10 mA)	120

		9.9.7 TT systems	122
		9.9.8 IT systems	122
		9.9.9 RCDs	122
	9.10	Earth monitoring	122
	9.11	Proving continuity	124
		9.11.1 Radial circuits	124
		9.11.2 Ring final circuits	127

Chapter 10 Particular issues of earthing and bonding — 129

10.1	Clean earths		129
10.2	PME for caravan parks		130
10.3	Exterior semi-concealed gas meters		131
10.4	Small-scale embedded generators		133
	10.4.1	Statutory regulations	133
	10.4.2	Engineering Recommendation G83	133
	10.4.3	Means of isolation	134
	10.4.4	Warning notices	135
	10.4.5	Up-to-date information	135
	10.4.6	The Stirling engine	136
	10.4.7	Guidance Note 7: *Special Locations*	136
10.5	Mobile and transportable units		136
	10.5.1	The term 'mobile or transportable unit'	136
	10.5.2	Examples of mobile or transportable units	137
	10.5.3	The risks	137
	10.5.4	Reduction of risks	137
	10.5.5	Supplies	138
	10.5.6	TN-C-S with PME	138
	10.5.7	Protection against direct contact	139
	10.5.8	Protection against indirect contact	139
10.6	Highway power supplies and street-located equipment		139
	10.6.1	Street furniture and street-located equipment	139
	10.6.2	Street furniture access doors	140
	10.6.3	Earthing of street furniture access doors	140
	10.6.4	Earthing of Class 1 equipment within street furniture and street-located equipment	140
	10.6.5	Distribution circuits	140
	10.6.6	Temporary supplies	140
10.7	Suspended ceilings		141
10.8	Exhibitions, shows and stands		143
	10.8.1	Protection by automatic disconnection of supply	143
	10.8.2	Distribution circuits	143
	10.8.3	Installations incorporating a generator	143
	10.8.4	Final circuits	144
10.9	Potentially explosive atmospheres		144
10.10	Medical locations		146
	10.10.1	TN-C system	146
	10.10.2	SELV and PELV	146
	10.10.3	Protection against indirect contact	146
	10.10.4	TT systems	147
	10.10.5	Medical IT systems	147
	10.10.6	Transformers for a medical IT system	147
	10.10.7	Supplementary equipotential bonding	148

10.11	Marinas	149
	10.11.1 The risks	149
	10.11.2 Minimising the risks	149
	10.11.3 Protection against direct contact	150
	10.11.4 Protection against indirect contact	150
	10.11.5 Isolating transformers	150
10.12	Cable tray and cable basket	151
10.13	Inspection and testing of protective conductors	153

Appendix A: Values for *k* for various forms of protective conductor 155

Appendix B: Data for armoured cables 157

Index 161

Cooperating organisations

The Institution of Engineering and Technology acknowledges the contribution made by the following representatives of organisations in the preparation of this publication:

BCA (British Cables Association)
C. Reed

BEAMA (The British Electrotechnical and Allied Manufacturers Association)
P. Galbraith, R. Lewington, M. Mullins, P. Sayer

Benchmark Electrical Safety Technology Ltd
G. Stokes

CORGI
P. Collins

Department of Trade and Industry
G. Scott, D. Tee

Electrical Contractors' Association
D. Locke, R. Lovegrove, L. Markwell

Electrical Contractors' Association of Scotland
D. Millar

Electrical Safety Council
M. Clark

Health and Safety Executive
K. Morton

Institution of Engineering and Technology
M. Coles, G. Cronshaw, N. Friswell, J. Hughes, D. Start, J. Ware

West Green Associates
G. Willard

Acknowledgements

References to British Standards are made with the kind permission of BSI. Complete copies can be obtained by post from:

BSI Customer Services
389 Chiswick High Road
London W4 4AL
Tel. General Switchboard: 020 8996 9000
Tel. Orders: 020 8996 9001
Fax. Orders: 020 8996 7001

Advice is included from Engineering Recommendation G83: *Recommendations for the connection of small-scale embedded generators (up to 16 A per phase) in parallel with public low voltage distribution networks* published by the Energy Networks Association. Complete copies of this and other Engineering Recommendations can be obtained by post from:

Energy Networks Association
18 Stanhope Place
Marble Arch
London
W2 2HH

This book was written by Eur Ing Geoffrey Stokes, BSc(Hons) C Eng, FIEE, FCIBSE
Managing Director of Benchmark Electrical Safety Technology Ltd
www.benchmarkelectrical.co.uk

Preface

This Guidance Note is part of a series issued by the Institution of Electrical Engineers (now the Institution of Engineering and Technology) to enlarge upon and simplify some of the requirements in BS 7671: *2001 Requirements for electrical installations* (also entitled the *IEE Wiring Regulations*, 16th Edition), as amended by Amendment No 2: 2004.

This Guidance Note does not ensure compliance with BS 7671. It is merely intended to explain some of the requirements of BS 7671 relating to earthing and bonding together with automatic disconnection. Readers should always consult the full text of BS 7671 to satisfy themselves of compliance and must rely upon their own skill and judgment when making use of the guidance provided here.

Electrical installations in the United Kingdom constructed to meet the requirements of BS 7671 are likely to satisfy the relevant aspects of statutory regulations such as the *Electricity at Work Regulations 1989*, but this cannot be guaranteed. It is stressed that it is essential to establish which statutory and other regulations apply and to install accordingly. For example, an installation in premises subject to licensing may have requirements differing from or additional to BS 7671, and these will take precedence. Users of this Guidance Note should also assure themselves that they have complied with any legislation that post-dates the publication.

Having been constructed to comply with BS 7671, the *Electricity at Work Regulations 1989* imposes a duty and responsibility on the owner/user of the electrical installation to maintain it so as to prevent danger.

Introduction

This Guidance Note is principally concerned with the electrical installation aspects of earthing and bonding together with automatic disconnection which forms the principal protective measure against indirect contact electric shock. It draws from the requirements embodied in the relevant parts, chapters and sections of BS 7671: 2001 *Requirements for Electrical Installations* (also entitled the *IEE Wiring Regulations,* 16th Edition), as amended by Amendment No 2: 2004. Where appropriate, other standards such as BS 7430: *Code of practice for earthing* are also referred to as well as international and European standards where relevant.

Earthing and bonding together with automatic disconnection are essential aspects of electrical installation design, the requirements of which depend to a large extent on the system type. These aspects are discussed in general terms but, where they apply to special applications, the advice is given in more specific terms drawing on guidance published in other IEE Guidance Notes and in particular in IEE Guidance Note 7: *Special locations.*

The electrical installation designer may find it helpful to refer to the guidance given here, but he/she would be required to take into account all other necessary aspects of the design and in the process involve other interested parties, such as:

- the designer
- the installer
- the electricity distributor
- the installation owner and/or user
- the architect
- the fire prevention officer
- all regulatory authorities
- licensing authority
- the Health and Safety Executive
- the insurers
- the planning supervisor.

In producing the design, advice should be sought from the installation owner and/or user as to the intended use. Often, as in a speculative building, the intended use is unknown. The specification and/or the operational manual must set out the basis of use for which the installation is suitable.

Precise details of each item of equipment should be obtained from the manufacturer and/or supplier and compliance with appropriate standards confirmed. This is particularly important with regard to the protective conductor currents of individual items of current-using equipment and the effect of the accumulation of such currents.

The operational manual should include a description of how the system as installed is to operate and all commissioning records. The manual should also include manufacturers' technical data for all items of switchgear, luminaires, accessories, etc. and any special instructions that may be needed. Section 6 of the *Health and Safety at Work etc. Act 1974* is concerned with the provision of information.

Guidance on the preparation of technical manuals is given in BS 4884: *Technical manuals* and BS 4940: *Technical information on construction products and services*. The size and complexity of the installation will dictate the nature and extent of the manual. This Guidance Note does not attempt to follow the pattern of BS 7671. The order in which the aspects of earthing and bonding as well as automatic disconnection appear is not significant.

Protective earthing

1.1 Protective earthing

Protective earthing is effected by connecting the source, such as the distribution transformer or generator, with earth and earthing of the fixed and portable equipment of the electrical installation which it supplies.

1.2 Source earthing

Sometimes referred to as supply system earthing, source earthing is the provision of a connection between the source of energy and the general mass of earth (Earth). This is normally achieved by connecting the neutral point or star point of the secondary winding of the transformer with earth, via a source earth electrode. This applies whether the source is a distribution transformer or generator. Figure 1.1 illustrates source earthing through a low-impedance source earth electrode and this relates to all TN systems as well as TT systems. For IT systems, a high impedance is inserted between the neutral or star point and the source earth electrode.

Source earthing in further detail is beyond the scope of this Guidance Note.

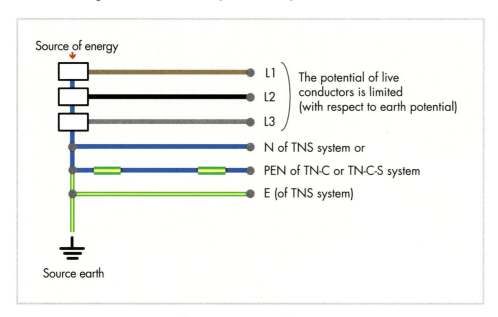

▶ **Figure 1.1**
Source earthing

1.3 The purpose of source earthing

The prime purpose of source earthing is to safeguard the security of the supply network by preventing the potential of the live conductors (with respect to earth) rising to a value inconsistent with their insulation rating which may otherwise fail to short-circuit with the consequential loss of supply.

For TT systems, the earthing of the source also provides an essential part of the path by which earth fault currents occurring in the installation flow back to the source. These earth fault currents can then be detected by the overcurrent protective devices (such as fuses or circuit-breakers) or residual current devices (RCDs) in order to automatically disconnect the faulty circuit and thus provide protection against indirect contact and, in the case of overcurrent protective devices, protection against fault currents.

For IT systems the source is connected to earth via a high impedance or is isolated electrically from earth.

1.4 The purpose of electrical equipment earthing

Electrical equipment earthing, or, alternatively, electrical installation earthing, is the connection of all exposed-conductive-parts of the installation to the intended means of earthing, and is always applicable where protection against indirect contact is by earthed equipotential bonding and automatic disconnection of supply (EEBADS). The purpose of such earthing is to facilitate the automatic operation of the protective devices for protection against electric shock and earth fault protection, so that the supply to the faulty circuit is disconnected promptly. Figures 1.2 and 1.3 show the earth fault current path of a circuit in which a line-to-earth fault has occurred in the installation, for a TN-C-S system and for a TT system, respectively.

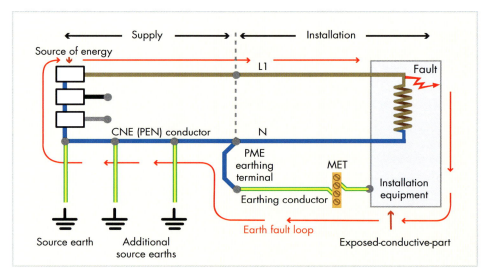

Figure 1.2 TN-C-S system – earth fault current path

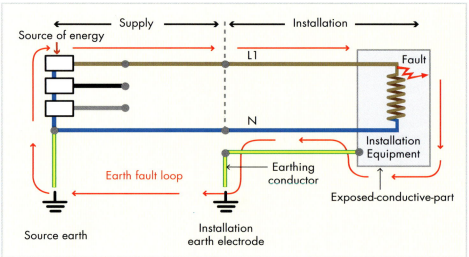

Figure 1.3 TT system – earth fault current path

The means of earthing

2.1 The means of earthing

The means of earthing is the arrangement which connects the general mass of Earth with the exposed-conductive-parts of an installation via the earthing conductor, the MET (main earthing terminal) and the circuit protective conductors, as shown in Figure 2.1.

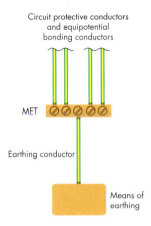

▶ **Figure 2.1**
The interconnections between the MET and exposed-conductive-parts

Essentially, the means of earthing can be of one of two forms:

▶ An earthing facility provided by the electricity distributor to the electricity consumer (e.g. for TN systems).
▶ An installation earth electrode provided by the consumer (e.g. for TT and IT systems).

For TN-S and TN-C-S systems, the means of earthing is normally provided by the electricity distributor, to which the consumer's earthing conductor is terminated. For TT and IT systems it falls to the consumer to provide such means of earthing.

For an installation forming part of a TT or IT system, the MET is connected to the installation earth electrode via the earthing conductor.

Regulation Group 542-02 of BS 7671: *Requirements for Electrical Installations* refers.

2.2 Distributors' facilities

For an installation forming part of a TN-S system, the MET is connected via the earthing conductor to the earthed star or neutral point of the source via the electricity distributor's supply cables, either by the cable armouring or sheath or by a separate conductor.

For an installation forming part of a TN-C system, a connection is required to be made between the MET and the electricity distributor's PEN conductor, via the earthing conductor.

For an installation forming part of a TN-C-S system, where the supply is PME (protective multiple earthing), the MET is connected to the earthed point of the source via the installation earthing conductor and the electricity distributor's PEN (protective earthed neutral) or CNE (combined neutral earth) conductor.

See also Clause 2.15: Responsibility for providing a means of earthing.

Regulations 542-01-01 to 542-01-04 refer.

2.3 Earth electrodes

It is sometimes necessary to provide the means of earthing for an installation by the use of an installation earth electrode, as in the case of an installation forming part of a TT system. Where an electricity distributor is unwilling or unable to provide a means of earthing in the form of an earth terminal connected to the supply cable, the installation owner is required to provide an installation earth electrode. Similarly, it may in certain circumstances be undesirable to employ an electricity distributor's earthing facility and the installation owner is then again required to provide an installation earth electrode.

There are various types of earth electrode:

- driven rods
- pipes
- tapes
- bare wires
- plates
- structural metalwork embedded in foundations
- lead cable sheath
- metal coverings of cables
- suitable metal reinforcement of concrete in reliable contact with earth
- structural steelwork of buildings in reliable contact with earth
- suitable underground metalwork.

2.4 Rod electrodes

Rods are typically constructed of solid copper, cylindrical in shape, but rods of copper molecularly bonded to steel are often used as can be rods made from galvanized steel or stainless steel. Where rigidity is necessary for driving, cruciform or star-shaped sections are sometimes preferred. Although more rigid for driving, these sections do not necessarily provide a significantly lower contact resistance, despite having a larger contact area.

The minimum generally recommended diameters for rods are 9, 12.5 and 15 mm for copper and copper-clad rods, and 16 mm for rods of galvanized steel and stainless steel. The preferred length of 9 mm diameter rods is 1.2 m and between 1.2 and 1.5 m for 12.5 and 15 mm diameter rods.

Table 4 of BS 7430: *Code of practice for earthing* gives guidance as to the minimum cross-sectional areas and diameters of various types of electrodes. The data is reproduced here as Table 2.1.

▶ **Table 2.1** Data from Table 4 of BS 7430: *Code of practice for Earthing*

Electrode type	Cross-sectional area (mm²)	Diameter or thickness (mm)
Copper strip	50	3
Hard drawn or annealed copper rods or solid wires for driving or laying in ground	50	8
Copper-clad or galvanised steel rods (see notes) for harder ground	153	14
Stranded copper	50	3 per strand

Notes:
1 For copper-clad steel rods the core should be of low-carbon steel with a tensile strength of approximately 600 N/mm² and a quality not inferior to grade S275 conforming to BS EN 10025. The cladding should be of 99.9% purity electrolytic copper, molecularly bonded to the steel core. The radial thickness of the copper should be not less than 0.25 mm.
2 Couplings for copper-clad steel rods should be made from copper-silicon alloy or aluminium bronze alloy with a minimum copper content of 75%.
3 For galvanized steel rods, steel of grade S275 conforming to BS EN 10025 should be used, the threads being cut before hot-dip galvanizing in accordance with BS 729.

An earth electrode consisting of a driven rod is suitable for providing a means of earthing for many, if not most, installations. However, a driven rod is not appropriate for applications in terrains containing hard strata such as rock.

Where deep driving of rods is necessary, a number of standard lengths of rod can be coupled together, as shown in Figure 2.2. As with driving a single rod, a drive cap should always be used and it is often less problematic to use an electric or pneumatic hammer with suitable rod adapter to avoid bending of the rod.

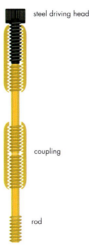

▶ **Figure 2.2** Rods coupled together

Regulation 542-02-04 of BS 7671 precludes the use of metalwork of a gas, water, oil or other service pipe to be used as an earth electrode.

The position of rod terminations, like all other electrode terminations, is required to be properly identified and accessible for testing and maintenance purposes, and Figure 2.3 illustrates a suitable arrangement for termination of the earthing conductor to the rod, consisting of:

▶ a rod type electrode
▶ a rod clamp
▶ a 'safety' notice
▶ an inspection pit with removable cover (preferably with the words 'Earth rod' embossed into the top surface).

Figure 2.3 Suitable arrangement for termination of the earthing conductor

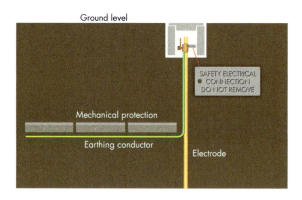

2.5 Tape and wire electrodes

Earth tapes and wires used as an electrode are typically made of untinned copper, strip or round section, and have a minimum cross-sectional area of 50 mm^2 (e.g. 12.5 x 4 mm). The minimum thickness of a tape is 3 mm and the minimum diameter for a strand of a stranded conductor is 3 mm. To avoid adverse soil resistivity conditions created by frosts, these electrodes should be laid at a depth of not less than 1 m. Figure 2.4 illustrates on the left a bare copper tape for use as an electrode buried in the ground, and on the right a suitable colour-identified insulated tape suitable for use as an earthing conductor to connect the electrode with the installation MET.

Figure 2.4 Bare and insulated copper tape

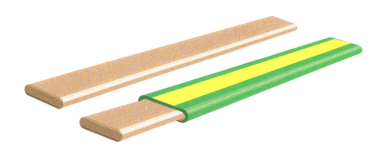

2.6 Plate electrodes

Earth plates can be made of copper or cast iron and are typically square with a surface area of 1 to 2 m^2 and are set vertically at a minimum depth of 600 mm from the ground surface to the top of the plate to ensure that the soil in close proximity is sufficiently damp. For rock occurring naturally near the surface, this minimum depth requirement may be relaxed. Figure 2.5 illustrates a vertically oriented earth plate and Figure 2.6 shows the connection of the earthing conductor to the plate.

Figure 2.5 A vertically oriented earth plate

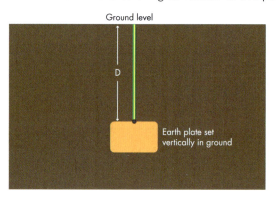

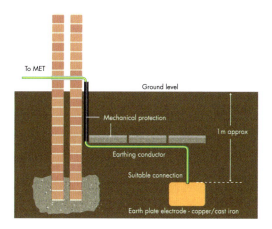

▶ **Figure 2.6** The connection of the earthing conductor to the plate

2.7 Structural metalwork electrodes

BS 7430: *Code of practice for Earthing* recognises that metalwork in foundation concrete can provide a convenient and effective earth electrode. Such electrodes, because of the large electrode area formed by the underground metalwork, can give an overall value of resistance to Earth of well below 1 Ω. However, there are issues that need to be taken into account before deciding as to the suitability of using this metalwork as an electrode.

One matter requiring consideration is the possibility of electrolysis and the consequential degradation of the metal. This may occur where the foundation metalwork is electrically in contact with, or bonded to, dissimilar buried metalwork which may result in corrosion of the metalwork and cracking of the surrounding concrete.

Where significant continuous earth leakage current containing a d.c. component exists, Clause 12.2 of BS 7430 recommends that an auxiliary earth electrode is bonded to the foundation metalwork.

Where the electrode is made up of structural steel, electrical continuity between all metalwork considered to be part of the earth electrode is essential. For electrical contacts between metalwork within concrete or below ground, such as between reinforcing bars, it is important to ensure this continuity by welding; above ground, joints may be made by attaching a bonding conductor to bypass each structural joint. This particularly applies to surfaces primed with paint before assembly.

BS 7430: *Code of practice for Earthing* also gives guidance on methods to calculate or measure the resistance to Earth of the arrangement and warns that these measured or calculated resistance values may increase over the lifetime of the installation.

Steel stanchions embedded in concrete in the ground may serve as suitable electrodes. Where thus used, it may be necessary to interconnect a number of such stanchions to provide for effective and long-term reliability.

Where the welded metal reinforcement grids in structural concrete are considered for use as an earth electrode, their use for such a purpose is required to have the prior agreement of the structural engineer.

The drilling of structural steelwork for connection of an earthing conductor, as shown in Figure 2.7, would require the consent of the structural engineer responsible for the design of the building structural components.

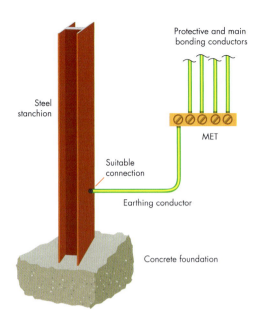

Figure 2.7 Earthing conductor connection to a structural steel stanchion

2.8 Metal covering of cables for electrodes

The lead sheath or other metal covering of a cable may be used as an earth electrode subject to the following conditions, as required by Regulation 542-02-05:

- adequate precautions are required to be taken to prevent excessive deterioration by corrosion, and
- the sheath or covering is required to be in effective contact with Earth, and
- the consent of the owner of the cable is required to be obtained, and
- arrangements are required to exist for the owner of the electrical installation to be warned of any proposed change to the cable which might affect its suitability as an earth electrode.

2.9 Location of the installation earth electrode

The location of the installation earth electrode will depend on a number of factors which the installation designer will need to take into account when siting the electrode(s), such as:

- the type of electrode
- the ground conditions
- the level of the water table and its seasonal variation
- the level at which soil with a suitable conductivity is found.

2.10 Resistance of the earth electrode

The electrical installation designer should determine the maximum acceptable resistance of the electrode and this may influence the choice of the type of electrode and/or its location.

In the case of rods and plates, in circumstances in which it may be necessary to achieve a low resistance, a number of rods or plates etc. can be connected in parallel. Dimensions and depths of the electrode, together with the resistivity of the soil, will greatly influence the resistance to Earth.

There are a number of factors which may adversely affect the resistance of an earth electrode, including:

- electrode contact resistance
- adjacent soil resistivity
- the size and shape of the earth electrode
- the burial depth.

The resistance between the electrode and any material in contact with it can be significantly increased by, for example:

- corrosion on the electrode
- coatings on the electrode, such as paint
- poor soil (e.g. builders' rubble)
- uncompacted or loose soil surrounding the electrode.

Soil resistivity is defined as a measure of the resistance of a cubic metre of soil, in units of ohm metres (Ωm), and is mainly dependent on soil composition, soil water content and the temperature of the soil. Soil resistivity is given the symbol rho, ρ.

Soil composition can vary enormously and be a mixture of sand, soil, chalk and rock to name just a few examples. Often the choice of position to site an electrode is very limited. Some soils are better than others and often the best results are obtained in damp and wet sand, peat clay, arable land, and wet marshy ground.

As Figure 2.8 illustrates, the resistivity can be significantly affected by the soil's moisture content. For example, according to the illustration, when the percentage moisture in the soil is reduced from 20 to 10 per cent, the corresponding soil resistivity increases dramatically.

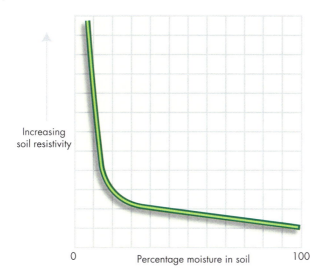

▶ **Figure 2.8** Example of the variation of soil resistivity with the moisture content.

The temperature of the soil does have an effect on the resistivity but this is marginal except at temperatures around 0 °C (the freezing point of water) where soil resistivity increases disproportionately.

BS 7430 provides data on soil resistivity for general guidance which is reproduced here for convenience in Table 2.2. Most of the United Kingdom would be represented in the second and third columns, but the values given should always be checked against particular local conditions.

Table 2.2 Soil resistivity (Ωm)

Type of soil	Climatic condition			
	Normal and high rainfall (i.e. greater than 500 mm a year)		Low rainfall and desert conditions (i.e. less than 250 mm a year)	Underground waters (saline)
	Probable value encountered	Range of values encountered	Range of values encountered	Range of values encountered
Alluvium and lighter clays	5	See note	See note	1 to 5
Clays (excluding alluvium)	10	5 to 20	10 to 100	1 to 5
Marls (e.g. Keuper marl)	20	10 to 30	50 to 300	
Porous limestone (e.g. chalk)	50	30 to 100		
Porous sandstone (e.g. Keuper sandstone and clay shales)	100	30 to 300		
Quartzites, compact and crystalline limestone (e.g. carboniferous sediments, marble, etc)	300	100 to 1000		
Clay slates and slatey shales	1000	300 to 3000	1000 upwards	30 to 100
Fissile slates, schists, gneiss and igneous rocks	2000	1000 upwards		

Note: Depends on water level of locality.

Not surprisingly, the size and shape of the electrode affects the resistance to Earth. Figure 4.9 replicates Figure 3 of BS 7430: *1998 Code of practice for Earthing* and gives data of the effects of rod diameter and rod length on electrode resistance for soil resistivity of 100 Ωm, assumed uniform.

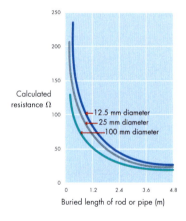

Figure 2.9 Variations in rod resistance with size and depth

2.10.1 Calculating the resistance of a rod electrode

Where it is necessary to predict the resistance to Earth of a vertical rod (or pipe) Equation (2.1) should be used:

$$R = \frac{\rho}{2\pi L}\left[\ln\left(\frac{8L}{d}\right) - 1\right] \; (\Omega) \qquad (2.1)$$

where: L is the length of the electrode in metres
d is the diameter of the electrode in metres
ρ is the resistivity of the soil (assumed uniform) in ohm metres
'ln' is the natural logarithm (log to the base 'e')

As a practical example of applying Equation (2.1), assume that the limiting value of the electrode resistance R is 50 Ω. Checking the resistance for a 15 mm diameter rod electrode buried in porous sandstone ground with a resistivity of 120 Ωm to a covered depth of 1.8 m and inserting the corresponding values into Equation (2.1), we get:

$$R = \frac{\rho}{2\pi L}\left[\ln\left(\frac{8L}{d}\right) - 1\right] = \frac{120}{2\pi \times 1.8}\left[\ln\left(\frac{8 \times 1.8}{0.015}\right) - 1\right] = 10.61[(\ln 960) - 1] = 10.61(6.87 - 1) \approx 62\,\Omega \quad (2.2)$$

Equation (2.2) predicts that a 15 mm diameter rod electrode buried to a depth of 1.8 m will not produce a low enough resistance (we are looking for a maximum of 50 Ωm). The diameter of the rod or its length can be changed so as to reduce the predicted resistance value. In this further example, we use a rod of twice the buried length of the original one (i.e. 3.6 m). Equation (2.3) gives the result as 35 Ω, which meets our requirements although two separate rods in parallel and sufficiently spaced may have produced an even better result:

$$R = \frac{\rho}{2\pi L}\left[\ln\left(\frac{8L}{d}\right) - 1\right] = \frac{120}{2\pi \times 3.6}\left[\ln\left(\frac{8 \times 3.6}{0.015}\right) - 1\right] = 5.31[(\ln 1920) - 1] = 5.31(7.56 - 1) \approx 35\,\Omega \quad (2.3)$$

2.10.2 Calculating the resistance of a plate electrode

For an electrode formed by a plate, the resistance to Earth, R, can be predicted by applying Equation (2.4):

$$R = \frac{\rho}{4}\sqrt{\frac{\pi}{2A}} \quad (\Omega) \quad (2.4)$$

where: ρ is the resistivity of the soil in ohm metres
A is the area of one face of the plate in square metres

2.10.3 Calculating the resistance of a tape electrode

For an electrode consisting either of tape or a bare round conductor the prediction of the resistance to Earth is more problematic, with the results depending on a number of further factors in addition to the length and buried surface area, including:

- the conductor's shape (round or flat)
- the conductor's arrangement or configuration in the ground.

BS 7430: *Code of practice for Earthing* gives comprehensive guidance on calculating resistances to Earth for these electrodes. Equation (2.5) gives one such method to predict the resistance of a solitary strip conductor run in a single straight line:

$$R = \frac{\rho}{2\pi L}\left[\ln\left(\frac{2L^2}{wh}\right) - 1\right] \quad (\Omega) \quad (2.5)$$

where: L is the length of the strip conductor in metres
h is the buried depth of the electrode in metres
w is the width of the strip in metres
ρ is the resistivity of the soil (assumed uniform) in ohm metres
'ln' is the natural logarithm (log to the base 'e')

2.10.4 Limiting values of the electrode resistance for TT systems

Where the electrode is provided for an installation that forms part of a TT system, and EEBADS is provided by an RCD, Regulation 413-02-20 requires that Equation (2.6) be satisfied:

$$R_A I_a \leq 50\,\text{V} \quad (2.6)$$

where: R_A is the sum of the resistances of the earth electrode and the protective conductor(s) connecting it to the exposed-conductive-part in ohms, and

I_a is the current causing the automatic operation of the protective device in 5 s. It should be noted that where an RCD device is used to provide automatic disconnection, then I_a is the rated residual operating current $I_{\Delta n}$.

For some special locations, Regulation 413-02-20 is modified so as to require the voltage in (2.6) not to exceed 25 V.

Equation (2.6) is generally easily satisfied for high-sensitivity RCDs. For example for a residual current device with a rated residual current of 30 mA, the limit on R_A would be 1666 Ω, as indicated in Equation (2.7):

$$R_A \leq \frac{50}{30 \times 1000} \leq 1666 \, \Omega \qquad (2.7)$$

Clearly, satisfying Equation (2.7) is only one consideration in choosing a limit on the earth electrode resistance. The installation designer needs to consider the stability of the electrode over the lifetime of the installation and, for example, the effects of variations in the water table level. BS 7430: *Code of practice for Earthing* recommends that the electrode resistance should not exceed 100 Ω, since it may otherwise become unstable.

2.11 Electrode installation

The design and construction of an earth electrode requires very careful consideration by the installation designer, as required by Regulation 542-02-03. Examples of issues likely to occur for most, if not all, electrode installations are:

- Damage – account to be taken of all potential causes of damage to which the earth electrode and earthing conductor could be subjected. Such causes may include, for example, vandalism, gardening, farm equipment, livestock animals, or future excavation associated with a building extension, and will depend to a large extent on the particular circumstances of the location.
- Corrosion – compatibility with the soil: copper is generally considered to be one of the better and more commonly used materials for earth electrodes. However, the corrosive effects of dissolved salts, organic acids and acid soils should be considered.
- Corrosion – galvanic effects: this may occur where items of dissimilar buried metalwork are electrically connected together. This electrolytic corrosion can have adverse effects on earth electrodes and earthing conductors as well as other underground services and structural metalwork. Table 2.3 below replicates data given in BS 7430: *Code of practice for Earthing* for the suitability of materials for bonding together.

▶ **Table 2.3** Suitability of materials for bonding together

Material assumed to have larger surface area	Electrode material or item assumed to have the smaller surface area			
	Steel	Galvanised steel	Copper	Tinned copper
Galvanised steel	Suitable	Suitable	Suitable	Suitable
Steel in concrete	Unsuitable	Unsuitable	Suitable	Suitable
Galvanised steel in concrete	Suitable	Suitable[1]	Suitable	Suitable
Lead	Suitable	Suitable[1]	Suitable	Suitable

[1] The galvanising on the smaller surface may suffer.

Where it is necessary to use a number of rods or other types of electrodes, care should be taken to position the electrodes so that they occupy positions which are not influenced by an adjacent electrode. A commonly adopted principle is to space adjacent electrodes at a distance not less than the buried depth of the electrode, as shown in Figure 2.10 where *d* represents the buried depth and *s* represents the minimum separation distance.

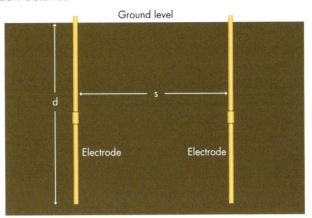

▶ **Figure 2.10** Separation distance for electrodes. *s* should be greater than *d*

2.12 Electrode loading capacity

The loading capacity of an earth electrode is dependent on:

▶ its shape
▶ its dimensions
▶ the current the electrode is required to carry
▶ the electrical and thermal properties of the soil.

Essentially, under all operating conditions, the heating effect due to energy dissipated into the soil will not normally result in a detrimental rise in the resistance of, or failure of, the earth electrode. The energy dissipated is I^2t, where *I* is the electrode current (amperes) and *t* is the duration (seconds). In this consideration, the designer would be required to take account of all currents, not just earth fault currents, likely to flow, such as functional currents.

Although the designer still needs to consider electrode load capacity, for a TT system where protection against indirect contact is provided by an RCD which provides automatic disconnection of the supply, the loading capacity requirement is generally likely to be met routinely.

2.13 Measurement of the electrode resistance

There are two basic methods for measuring the resistance of an earth electrode, namely by the use of a proprietary test instrument designed for the purpose, or by the use of an earth fault loop impedance test instrument. Both methods are described here.

2.13.1 Proprietary earth electrode test instrument

Before a test on an existing installation earth electrode using a proprietary earth electrode test instrument (method 1) is undertaken, it is essential to ensure that the installation is securely isolated from the supply. It is also necessary to disconnect the earthing conductor from the earth electrode. This disconnection will ensure that the test current only passes through the earth electrode and not through parallel paths. The installation is required to remain isolated from the supply until all testing has been completed and the earth electrode connection reinstated.

Ideally, the test should be carried out when the ground conditions are least favourable, such as during dry weather.

The test requires the use of two temporary test spikes (electrodes), and is carried out in the following manner.

Connection to the earth electrode is made using terminals C1 and P1 of a four-terminal earth tester. To exclude the resistance of the test leads from the resistance reading, individual leads should be taken from these terminals and connected separately to the electrode. Where the test lead resistance is insignificant, the two terminals may be short-circuited at the test instrument and connection made with a single test lead, the same being true if using a three-terminal tester. Connection to the temporary spikes is made as shown in Figure 2.11.

The distance between the test spikes is important. Where they are too close together, their resistance areas will overlap. In general, reliable results may be expected if the distance between the electrode under test and the current spike, C2, is at least ten times the maximum dimension of the electrode system, e.g. 30 m for a 3 m long rod electrode.

Three readings are taken:

- with the potential spike, P2, initially midway between the electrode and current spike,
- secondly at a position 10 per cent of the electrode-to-current spike distance back towards the electrode, and
- finally at a position 10 per cent of the distance towards the current spike.

By comparing the three readings, a percentage deviation can be determined. This is calculated by taking the average of the three readings, finding the maximum deviation of the readings from this average in ohms, and expressing this as a percentage of the average.

The accuracy of the measurement using this technique is typically 1.2 times the percentage deviation of the readings. It is difficult to achieve an accuracy of measurement better than 2 per cent, and inadvisable to accept readings that differ by more than 5 per cent. To improve the accuracy of the measurement to acceptable levels, the test is required to be repeated with a larger separation between the electrode and the current spike.

▶ **Figure 2.11** Earth electrode test

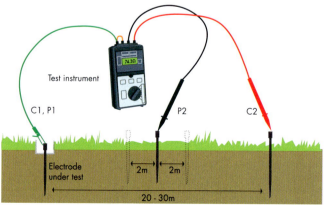

The test instrument output may be a.c. or reversed d.c. to overcome electrolytic effects. Because these instruments employ phase-sensitive detectors, the errors associated with stray currents are eliminated.

The instrument should be capable of checking that the resistance of the temporary spikes used for testing is within the accuracy limits stated in the instrument specification. This may be achieved by an indicator provided on the instrument, or the instrument should have a sufficiently high upper range to enable a discrete test to be performed on the spikes.

Where the resistance of the temporary spike is too high, measures to reduce the resistance will be necessary, such as driving the spikes deeper into the ground or watering with brine to improve the contact resistance. In no circumstances should these techniques be used to temporarily reduce the resistance of the earth electrode under test.

On completion of the testing procedure always ensure that the earthing conductor is properly reconnected.

2.13.2 Earth fault loop impedance test instrument for measuring electrode resistance

For the case where the electrode under test is the means of earthing for an installation forming part of a TT system, an alternative method of electrode resistance measurement may be used (method 2). In such cases, automatic disconnection for protection against indirect contact is likely to be by an RCD and precision in the measurement is not absolutely necessary.

Before a test on an installation earth electrode using method 2 is undertaken it is essential to ensure that the installation is securely isolated from the supply. It is then necessary to disconnect the earthing conductor to the earth electrode. This disconnection will ensure that all the test current passes through the earth electrode alone and not through parallel paths. The installation is required to remain isolated from the supply until all testing has been completed and the earth electrode connection reinstated.

An earth fault loop impedance test instrument is connected between the phase conductor at the source of the installation and the earth electrode, and a test performed. Because the resistance to Earth of the electrode is likely to far exceed the contribution made by other constituent parts of the loop, the impedance reading taken is treated as the electrode resistance.

Regulation 413-02-20 requires that the electrode resistance, R_A, does not exceed that given by Equation (2.8) for installations where dry conditions exist. For some special locations, such as construction sites and agricultural and horticultural premises, Equation (2.9) applies (see Regulations 604-05-01 and 605-06-01):

$$R_A I_{\Delta n} \leq 50 \text{ V} \tag{2.8}$$

$$R_A I_{\Delta n} \leq 25 \text{ V} \tag{2.9}$$

where: R_A is the sum of the resistances of the earth electrode and the protective conductor(s) connecting it to the exposed-conductive-parts, and
$I_{\Delta n}$ is the rated residual operating current (A).

Maximum values of R_A for the basic standard ratings of RCDs are given in Table 2.4, unless the manufacturer declares alternative values.

▶ **Table 2.4** Maximum permitted values of earth electrode resistance, R_A, for installations forming part of a TT system

Rating of the RCD (mA)	Maximum permitted value of electrode resistance, R_A	
	Dry conditions (Ω)	Construction sites, agricultural and horticultural premises (Ω)
30	1666	833
100	500	250
300	160	80
500	100	50

The table indicates that the use of a suitably rated RCD will theoretically allow much higher values of R_A, and therefore of the total earth fault loop impedance, Z_s, than could be expected by using the overcurrent devices for protection against indirect contact. However, values of electrode resistance exceeding 100 Ω would require further investigation in order for the installation designer to be satisfied that the electrode resistance will remain stable over the lifetime of the installation.

On completion of the testing procedure it should be ensured that the earthing conductor is properly reconnected.

2.14 Determination of the external earth fault loop impedance, Z_e

The external earth fault loop impedance, Z_e, is part of the total earth fault loop impedance and it is essential that an ohmic value for it is determined. There are three ways that the installation designer is permitted to determine this value:

▶ by enquiry to the electricity distributor
▶ by calculation
▶ by measurement.

(Regulation 713-11-01 refers.)

Generally before commencing a detailed design, the installation designer would seek details of the supply characteristics from the electricity distributor, which would include maximum values for prospective fault levels and for the external earth fault loop impedance, Z_e.

For the case where a low voltage distribution transformer is owned by the consumer, or where such a transformer is dedicated to supply that customer, the value both for prospective fault level and for the external earth fault loop impedance may be calculated. However, a detailed knowledge of the constituent parts of the loop is required and this can sometimes present difficulties.

Even where the external earth fault loop impedance has been determined, either by enquiry or calculation, it is still necessary for this parameter to be measured in order to ensure that:

- The means of earthing is properly connected to the source star or neutral point (either by a metal conductor in the case of TN systems or by an installation earth electrode in the case of TT and IT systems).
- The measured value does not exceed the designer's expected value.

The measurement of the external earth fault loop impedance, Z_e, is made between the phase conductor of the supply and the means of earthing, with the main switch open and with all the circuits securely isolated from the supply. The means of earthing is required to be disconnected from the installation's earthed equipotential bonding for the duration of the test to remove parallel paths. Care should be taken to avoid any shock hazard to the testing personnel and other persons on the site, both whilst establishing contact and performing the test.

On completion of the measurement procedure, ensure that the earth connection has been replaced before re-closing the main switch.

(Regulations 542-04-02 and 711-01-01 refer.)

2.15 Responsibility for providing a means of earthing

Many, if not most, electrical installations are required to be earthed for safety reasons. In fact, under the *The Electricity Safety, Quality and Continuity Regulations 2002* (*ESQCR*) the consumer is responsible for ensuring that the installation is satisfactorily earthed. *ESQCR* Regulations 26(1) and 26(2) require the consumer's installation to be 'so constructed, installed, protected and used or arranged for use so as to prevent, so far as is reasonably practicable, danger or interference with the distributor's network or with supplies to others'. A consumer's installation complying with the requirements of BS 7671 is deemed to meet these statutory requirements.

Under Regulation 24(4) of the *ESQCR*, the electricity distributor is generally obliged to provide the consumer with an earthing facility for a new supply to a low voltage installation. The electricity distributor is excused from this obligation if there are safety reasons which preclude the provision of an earthing facility.

Where an existing installation is supplied from the public low voltage network, the electricity distributor is not obliged to provide an earthing terminal. However, the electricity distributor may be willing to do so.

Where the electricity distributor does provide an earthing facility, the responsibility of ensuring the safety and efficacy of this facility rests with the electricity distributor, as required by Regulation 24(1) of the *ESQCR*. However, it is for the consumer, or his/her agent, to make certain that the earthing facility is suitable for the requirements of the electrical installation and that it is properly connected to the MET of the installation.

Where an earthing facility is not provided by the electricity distributor it should never be assumed that, because a supply cable metal sheath appears to be earthed, such a facility is available. Under no circumstances should the consumer, or his/her agent, connect to the sheath of a supply cable. Such an unauthorized practice is both illegal and dangerous.

The earthing conductor 3

3.1 The earthing conductor

The earthing conductor of an electrical installation is the protective conductor that connects the means of earthing with the MET of the installation, as illustrated in Figure 3.1. There is only one such conductor in each installation.

Typically, and for practical considerations, the earthing conductor is provided by:

- a single-core cable, or
- a core of a multi-core cable, or
- a strip of tape conductor.

Not commonly utilized, other types of conductor are permitted by BS 7671, such as:

- metal sheath of a cable, or
- metal armouring of a cable, or
- metal conduit.

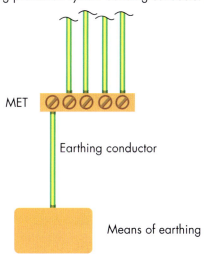

▶ **Figure 3.1** Conceptual layout of the earthing conductor and the means of earthing

A means should be provided to facilitate the disconnection of the earthing conductor in order to measure the external earth fault loop impedance; this should be achieved in an accessible position. The means of disconnection may be in the form of a disconnectable link combined with the MET, as shown in Figure 3.2, or may be a joint capable of disconnection only by means of a tool (e.g. spanner or screwdriver).

(Regulation 542-04-02 refers.)

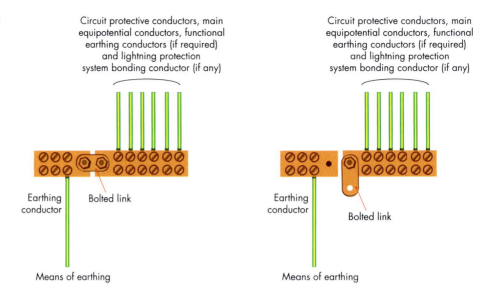

Figure 3.2 Illustration of an example of a disconnectable link

3.2 The cross-sectional area of an earthing conductor

Subject to certain limits, the minimum cross-sectional area (CSA) of an earthing conductor is determined in one of two ways:

- calculated using the adiabatic equation given in Regulation 543-01-03 (reproduced below for convenience as Equation (3.1)), or
- by referring to Table 54G given in Regulation 543-01-04 (reproduced for convenience as Table 3.1).

$$S \geq \frac{\sqrt{(I^2 t)}}{k} \ (\text{mm}^2) \tag{3.1}$$

where: S is the nominal cross-sectional area of the conductor in mm².

I is the value in amperes (rms for a.c.) of the fault current for a fault of negligible impedance, which can flow through the associated protective device, due account being taken of the current limiting effect of the circuit impedances and the limiting capability ($I^2 t$) of that protective device. Note that account needs to be taken of the effect, on the resistance of circuit conductors, of their temperature rise as a result of overcurrent (see Regulation 413-02-05).

t is the operating time of the disconnecting device in seconds corresponding to the fault current I in amperes.

k is a factor that takes account of the resistivity, temperature coefficient and heat capacity of the conductor material, and the appropriate initial and final temperatures. Values of k for protective conductors in various use or service are as given in Tables 54B, 54C, 54D, 54E and 54F. The values are based on the initial and final temperatures indicated below each table.

Where the application of the formula produces a non-standard size, a conductor having the nearest larger standard cross-sectional area shall be used.

As an alternative to the use of Equation (3.1) to determine the CSA of the earthing conductor, Table 3.1 gives data for the selection of the CSA.

▶ **Table 3.1** Data from Table 54G of BS 7671. Minimum CSAs of protective conductors in relation to the CSAs of associated phase conductors

CSA of phase conductor S (mm²)	Minimum CSA of corresponding protective conductor (mm²)	
	If protective conductor is of the same material as the phase conductor (mm²)	If protective conductor is not of the same material as the phase conductor (mm²)
$S \leq 16$	S	$\frac{k_1}{k_2} \times S$
$16 < S \leq 35$	16	$\frac{k_1}{k_2} \times 16$
$S > 35$	$\frac{S}{2}$	$\frac{k_1}{k_2} \times \frac{S}{2}$

Notes:
1. k_1 is the value for the phase conductor, selected from Table 43A of BS 7671 (replicated in Appendix A) according to the materials of both conductor and insulation.
2. k_2 is the value of k for the protective conductor, selected from Tables 54B, 54C, 54D, 54E or 54F of BS 7671, as appropriate (replicated in Appendix A).

As for all protective conductors, an earthing conductor with a CSA of 10 mm² or less is required to be of copper (see Regulation 543-02-03).

It has to be acknowledged that the use of Equation (3.1) to determine the CSA of an earthing conductor is by far more problematic than to select the CSA by referring to Table 54G of BS 7671 (Table 3.1 in this Guide). The use of Equation (3.3) is rendered ineffective in the case of PME supplied installations where Regulation 542-03-01 requires the earthing conductor to also meet the CSA requirements of Regulation 547-02-01 for main equipotential bonding conductors. In other words, the CSA of the earthing conductor should be able to meet the requirements for a main bonding conductor given in Table 54 H of BS 7671, which for convenience is reproduced as Table 3.2.

▶ **Table 3.2** Minimum CSA of main equipotential bonding conductors in relation to the neutral of the supply

Copper equivalent CSA of the supply neutral conductor (mm²)	Minimum copper equivalent* CSA of the main equipotential bonding conductor (mm²)
< 35	10
35 – 50	16
50 – 95	25
95 – 150	35
> 150	50

* copper or copper equivalent (in conductance terms)

It should be noted that the minimum CSAs of main equipotential bonding conductors given in Table 54H of BS 7671 may be modified by the electricity distributor. In other words, the electricity distributor may require main equipotential bonding conductors with a larger CSA than those given in Table 54H, for example because of network conditions.

3.3 The CSA of a buried earthing conductor

For an earthing conductor buried in the ground, a further set of minimum CSAs are given in Table 54A of BS 7671, reproduced for convenience as Table 3.3.

▶ **Table 3.3** Minimum CSA of a buried earthing conductor

	Protected against mechanical damage	Not protected against mechanical damage
Protected against corrosion by a sheath	As required by Regulation 543-01	16 mm² copper 16 mm² coated steel
Not protected against corrosion	25 mm² copper 50 mm² steel	25 mm² copper 50 mm² steel

BS 7430: *Code of practice for earthing* recommends that for a tape or strip conductor, the thickness should be such as to withstand mechanical damage and corrosion.

3.4 Minimum CSA of an earthing conductor

Irrespective of the method used to determine the minimum CSA of an earthing conductor, Regulation 543-01-01 stipulates that for a copper earthing conductor which is not an integral part of a cable (such as the core of a cable) and is not contained in an enclosure (such as a conduit) formed by a wiring system, the required minimum CSA is:

▶ 2.5 mm² where protection against mechanical damage is provided,
▶ 4 mm² where such protection is not provided.

3.5 Impedance contribution of an earthing conductor

Very exceptionally, an earthing conductor may be of such length that its CSA may need to be increased to minimise the contribution it makes to the overall earth fault loop impedance (Z_s), so as to ensure protection against indirect contact for all circuits downstream. This impedance can be reduced by selecting a protective conductor of increased CSA or, for an earthing conductor contained within a composite cable, by increasing the CSAs of all conductors (including that of the line conductors).

3.6 Colour identification of an earthing conductor

If the earthing conductor is a single-core cable or the core of a cable it must be identified with the colour combination green and yellow, as required by Regulations 514-03-01 and 514-04-02. See Figure 3.1 for an example of this colour identification scheme in use.

Another example is that a tape, strip or other bare conductor where used as an earthing conductor must be identified at intervals with the colour combination green and yellow (see Regulation 514-04-06).

3.7 Protection of an earthing conductor against external influences

As with all other equipment, an earthing conductor and its electrical connections must be protected from mechanical damage (e.g. impact and vibration), corrosion (e.g. electrolysis) and other external influences to which they may be expected to be exposed. Additionally, an earthing conductor, as for all conductors, must be properly supported throughout its length, and particularly at the terminations.

Where an earthing conductor with a CSA of 6 mm² or less is not part of a multi-core cable and not enclosed by conduit or trunking, it is a requirement of Regulation 543-03-02 for it to be protected throughout by a covering at least equivalent to that of the insulation of a single-core non-sheathed cable of the appropriate size and having a voltage rating of at least 450/750 V.

Regulation 543-03-02 also requires that an uninsulated earthing conductor with a CSA of 6 mm² or less, that forms part of a cable, must be protected by insulating sleeving where the sheath of the cable is removed adjacent to joints and terminations.

As required by Regulation 543-03-01, in situations in which an earthing conductor is likely to corrode or be subject to mechanical damage its CSA may need to be larger than otherwise determined, in order to protect the conductor against these problems.

However, with regard to mechanical damage, other means of protection may be provided such as enclosing the earthing conductor in a conduit or in trunking. Where this method is adopted and the conduit or trunking is ferrous, it is important to ensure that the associated live conductors (both phase conductor(s) and neutral conductor) are also contained within the same conduit or trunking (see Regulation 521-02-01).

3.8 Disconnection of the earthing conductor

As previously mentioned, Regulation 542-04-02 requires a means for disconnection of the earthing conductor to be provided at or near the MET of an installation. Such a means is necessary to facilitate the measurement of the upstream earth fault loop impedance, Z_e. Additionally, it is required that the means of disconnection can only be effected by use of a tool and it also needs to be mechanically strong so as to ensure continuity at all times. Figure 3.2 illustrates a typical means for disconnection of the earthing conductor.

3.9 Connection of the earthing conductor to the means of earthing

Where the means of earthing is provided by the electricity distributor the connection of the earthing conductor is straightforward and needs no explanation.

However, where the means of earthing is an earth electrode, special care is required to ensure that the electrode is not subject to damage, corrosion or other deterioration. Guidance on these aspects is given in Chapter 2.

The connection of the earthing conductor to the means of earthing is required to be identified by the attachment of a label, as shown in Figure 3.3.

▶ **Figure 3.3** Label to be attached to the connection of the earthing conductor

System types and earthing arrangements

4.1 A system

In electrical installation terms, a system has two constituent parts, namely a single source of energy and an electrical installation. The systems are described in a vocabulary in which the various letters have particular meanings, such as those given in Table 4.1. The term system is defined in Part 2 of BS 7671.

▶ **Table 4.1** System type designation letters and their meanings

First letter	Second letter	Subsequent letters
Source earthing arrangements	Arrangement of connection of exposed-conductive-parts of the installation with earth	Arrangement of protective and neutral conductors
'T' Direct connection of source with Earth at one or more points (e.g. one pole of a single-phase source or the star point of a three-phase source)	'N' Exposed-conductive-parts of the installation connected directly by a protective conductor with the source earth	'C' – 'Combined' Single conductor provides both neutral and protective conductor functions
		'S' – 'Separate' Separate conductors for neutral and protective conductor functions
		'C-S' Neutral conductor and protective conductor combined in the supply and separate in the installation
	'T' Exposed-conductive-parts of the installation connected by protective conductor(s) to an independent earth electrode and via the conductive mass of the earth to the source earth	
'I' All source live parts isolated from Earth or connected by a high impedance to earth	'T' Exposed-conductive-parts of the installation connected by protective conductor(s) to an independent earth electrode and via the conductive mass of the earth to the source earth	

Note: The *ESQCR 2002* preclude the IT system for use on public network supplies.

As can be seen from Table 4.1, the first and second letters refer to the source earthing arrangements and the others relate to the installation neutral and protective conductor arrangements. There are five distinct system types:

- TN-C system
- TN-S system
- TN-C-S system
- TT system
- IT system.

Before commencing on the detail of the design of an electrical installation, it is important for the designer to determine the system type, particularly from the viewpoint of earthing and equipotential bonding as well as protection against indirect contact. For example, the system type will have a significant influence on the CSAs of the earthing and equipotential bonding conductors (see Regulation 312-03-01).

For the majority of cases, the source is provided by an electricity distributor whose responsibility it is to earth the source (e.g. distribution transformer). However, for consumers taking the supply at high voltage, and where the distribution transformer is owned by the consumer, the responsibility for earthing the source rests with the consumer. Similarly, it is the consumer's responsibility for earthing the source where the source is a privately owned generator.

Available guidance documents relating to source earthing are:

- BS 7430: *Code of practice for earthing*, published by the British Standards Institution.
- *Guidelines for the design, installation, testing and maintenance of main earthing systems in substations*, originally published by the Electricity Association and now available from the Energy Networks Association.
- The *ESQCR 2002* (Regulation 8 refers).

4.2 TN-C system

Figure 4.1 shows a TN-C system consisting of a single three-phase source and a three-phase electrical installation with three line conductors (three phase conductors) and a combined neutral and protective conductor throughout. The star point, or neutral if single-phase, of the source is earthed through a low-impedance earth electrode to the general mass of earth.

It is important to note that in a TN-C system the neutral and protective earth conductor functions are combined both in the supply and in the installation; in other words the PEN or CNE is combined throughout the system and the exposed-conductive-parts of the installation are connected by this conductor back to the source. This conductor provides a return path both for the neutral conductor current to flow under normal conditions and for the earth fault current to flow for the duration of a line-to-earth fault occurring in the installation. Figure 4.2 shows the complete path that fault current will flow in conditions of a line-to-earth fault occurring in the installation.

▶ **Figure 4.1** TN-C system (three-phase)

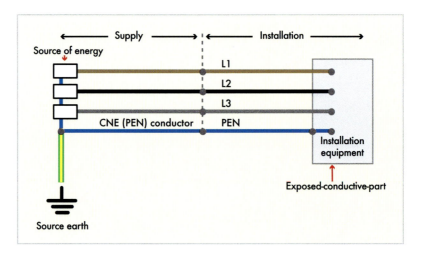

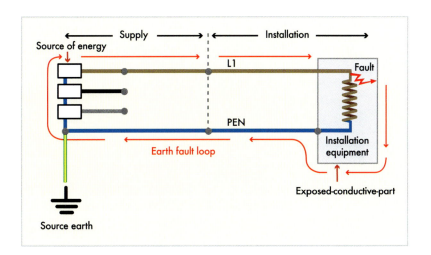

▶ **Figure 4.2** TN-C system showing an earth fault loop

This system type is very infrequently used in the United Kingdom. Regulation 546-02-01 of BS 7671 restricts the use of an installation PEN conductor to circumstances where:

▶ as required by the *ESQCR*, the consumer has obtained an exemption from the ban on CNE conductors in the installation, or
▶ the source is a privately owned transformer, private generating plant or other source and connected in such a manner that there is no electrical connection with the public distribution system.

4.3 TN-S system

Figure 4.3 shows a TN-S system consisting of a single three-phase source and a three-phase electrical installation with four live conductors (three phase conductors and a neutral conductor) and a protective conductor. The star point of the source is earthed through a low impedance earth electrode to the general mass of Earth. For a single-phase source the source neutral would be similarly earthed.

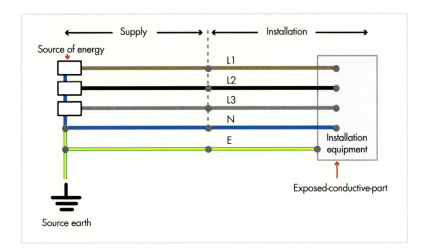

▶ **Figure 4.3** TN-S system (three phase)

▶ **Figure 4.4** TN-S system showing an earth fault loop

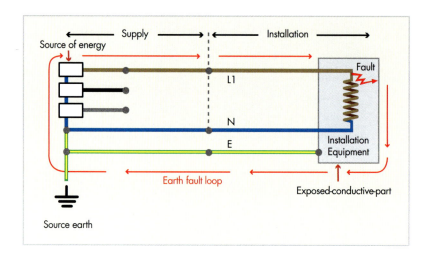

It is important to note that in a TN-S system the neutral conductor and the protective earth conductor functions are separate both in the supply and in the installation. The exposed-conductive-parts of the installation are connected by the protective earth conductor back to the source. This conductor provides a return path for earth fault current to flow for the duration of a line-to-earth fault occurring in the installation. Figure 4.4 shows the complete path that fault current will flow in conditions of a line-to-earth fault occurring in the installation.

This type of system will be commonly found in many systems supplied from the public supply networks which pre-date PME supplies. The electricity distributor often provides an earthing facility connected to the supply cable lead sheath or cable steel armouring or by utilising a separate supply protective conductor, terminated in a suitable earthing terminal located at the supply intake position.

It is worth noting that a supply to an installation forming part of a TN-S system can be provided by an overhead distribution system as well as by the more commonly encountered supply cable buried underground. Figure 4.5 illustrates the two forms of supply.

▶ **Figure 4.5** TN-S system (single phase) shown pictorially with an overhead and an underground supply cable, respectively

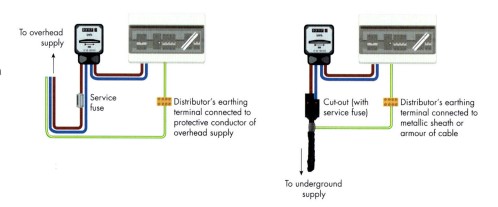

4.4 TN-C-S system

Figure 4.6 shows a TN-C-S system consisting of a single three-phase source and a three-phase electrical installation with three line conductors (three phase conductors) and a combined neutral and protective conductor in the supply. The star point of the source is earthed through a low-impedance earth electrode to the general mass of earth. For a single-phase source the source neutral would be similarly earthed.

It is important to note that in a TN-C-S system the neutral and protective earth conductor functions are combined in the supply and separate in the installation. The exposed-conductive-parts of the installation are connected by this separate protective conductor in the installation to the combined neutral and protective conductor of the supply back to the source. This installation protective conductor provides a return path for earth fault current to flow for the duration of a line-to-earth fault occurring in the installation. The combined neutral and protective conductor of the supply provides a return path both for neutral conductor current to flow under normal conditions and for earth fault current to flow for the duration of a line-to-earth fault occurring in the installation. Figure 4.7 shows the complete path that the fault current will flow under conditions of a line-to-earth fault occurring in the installation.

For TN-C-S systems with supplies from the public distribution network, the separation of the functions of neutral conductor and protective conductor occurs at the supply intake position. For privately owned sources, the transitional point is normally at the consumer's main switchgear.

There are two basic forms of TN-C-S systems, namely:

▶ TN-C-S system with PME
▶ TN-C-S system with PNB (protective neutral bonding).

For the TN-C-S system (PME) variant, as shown in Figure 4.6, the supply PEN or CNE is earthed at multiple points of the supply, as well as at the source itself, thereby providing a low-impedance path to Earth for all parts of the PEN conductor. This system has been used by electricity distributors, and their predecessors, for almost all new low voltage supplies installed since the 1970s.

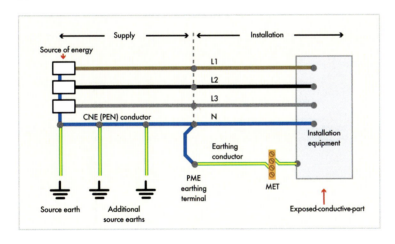

▶ **Figure 4.6** TN-C-S system (PME)

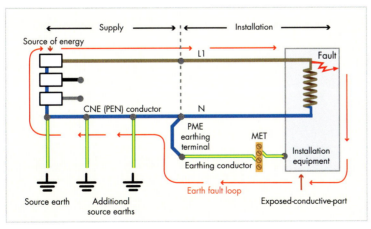

▶ **Figure 4.7** TN-C-S system (PME) showing an earth fault loop

It is worth noting that the supply to a TN-C-S system can be provided by an overhead distribution system as well as the more commonly used supply cable buried underground. Figure 4.8 illustrates the two forms of supply.

▶ **Figure 4.8** TN-C-S system (single-phase) shown pictorially with an overhead and an underground supply cable, respectively

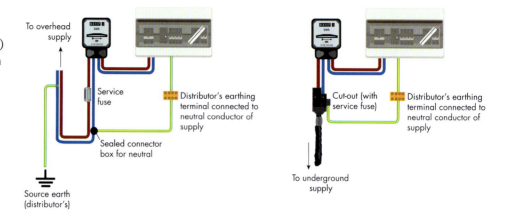

The use of the alternative TN-C-S system (PNB), as shown in Figure 4.9, is confined to cases in which a single consumer is fed from an electricity distributor's distribution transformer or other source. The PEN or CNE is connected to Earth at one point only usually close to the source, although this is not necessarily the case (see alternative positions shown in Figure 4.9).

For sources owned by the electricity distributor, the responsibility for earthing rests with the distributor. Alternatively, where the consumer owns the source, it is the consumer's responsibility to earth the source, and this is sometimes carried out at a location near the consumer's main switchgear position.

▶ **Figure 4.9** TN-C-S system (PNB)

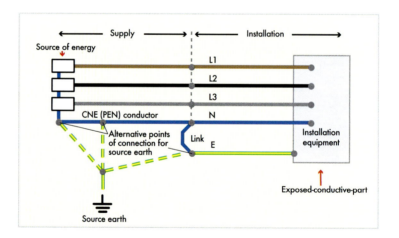

4.5 TT system

Figure 4.10 shows a TT system consisting of a single three-phase source and a three-phase electrical installation with three line conductors (three phase conductors) and a neutral conductor. The star point of the source is earthed through a low-impedance earth electrode to the general mass of Earth. For a single-phase source the source neutral would be similarly earthed.

It is important to note that in a TT system the exposed-conductive-parts of the installation are connected by a protective conductor in the installation to the MET and hence to the installation earth electrode which is electrically independent of the source earth. These components of the installation provide a path for the earth fault current to flow back to the source during a line-to-earth fault in the installation. Figure 4.11 shows the complete path that fault current will take in conditions of a line-to-earth fault occurring in the installation.

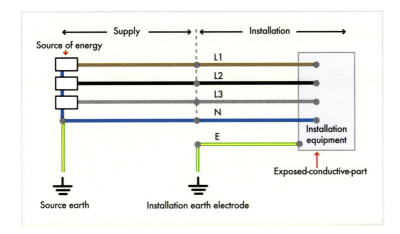

▶ **Figure 4.10** TT system

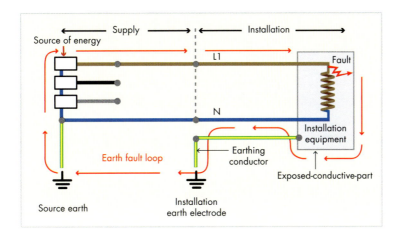

▶ **Figure 4.11** TT system showing an earth fault loop

It is important to recognise that the supply to an installation forming part of a TT system may include a protective conductor. However, for an installation to form part of a TT system the exposed-conductive-parts of the installation are required to be solely connected to the installation earth electrode.

The TT system is often used where a designer considers it undesirable to use the earthing facility offered by the electricity distributor where it is from a PME supply network. Alternatively in some cases, for statutory and/or technical reasons, the electricity distributor may be unwilling to offer an earthing facility and in this case a TT system provides a solution.

It is worth noting that a supply to an installation forming part of a TT system can be provided by an overhead distribution system as well as the more common one of a supply cable buried underground. Figure 4.12 illustrates the two forms of supply.

▶ **Figure 4.12** TT system (single-phase) shown pictorially with an overhead and an underground supply cable, respectively

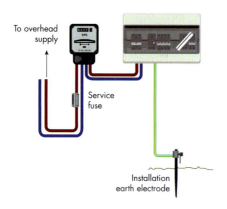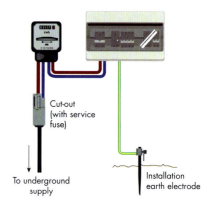

4.6 IT system

Figure 4.13 shows an IT system consisting of a single three-phase source and a three-phase electrical installation with three line conductors (three phase conductors) and a neutral conductor. The star point of the source is earthed through a low-impedance earth electrode to the general mass of earth via an earthing impedance. For a single-phase source the source neutral would be similarly earthed.

As the source is not earthed directly in an IT system, such a low voltage system is precluded from use in public distribution networks in the United Kingdom by the *ESQCR 2002*.

▶ **Figure 4.13** IT system

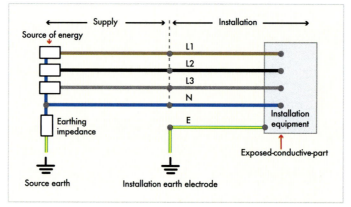

It is important to note that in an IT system the exposed-conductive-parts of the installation are connected by a protective conductor in the installation to the MET and hence to the installation earth electrode which is electrically independent of the source earth. These components of the installation provide a path for the earth fault current to flow back to the source during a line-to-earth fault in the installation. Figure 4.14 shows the complete path that fault current will take in conditions of a line-to-earth fault occurring in the installation.

▶ **Figure 4.14** IT system showing an earth fault loop

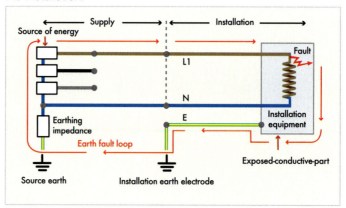

Main equipotential bonding 5

5.1 The purpose of main equipotential bonding

Main equipotential bonding is not to be confused with earthing. Bonding serves the function of minimising the magnitude of touch voltages within the building where an earth fault occurs in the installation.

Touch voltages occur when an earth fault develops in the installation. An earth fault current is defined as a fault current which flows to earth. As this current flows to Earth, touch voltages can be generated by the impedances between:

▶ exposed-conductive-parts and other exposed-conductive-parts, and
▶ extraneous-conductive-parts and other extraneous-conductive-parts, and
▶ exposed-conductive-parts and extraneous-conductive-parts, and
▶ exposed-conductive-parts and Earth.

As Figure 5.1 illustrates, an earth fault on circuit B produces a touch voltage between exposed-conductive-parts of that circuit and exposed-conductive-parts of circuit A and, in turn, to the extraneous-conductive-parts. Equation (5.1), can be used to predict the touch voltage U_t from the nominal voltage U_0, the total earth fault loop impedance Z_s and the resistance of the circuit protective conductor in circuit B, i.e. $R_{2(B)}$:

$$U_t = \frac{U_0}{Z_s} R_{2(B)} \quad (V) \tag{5.1}$$

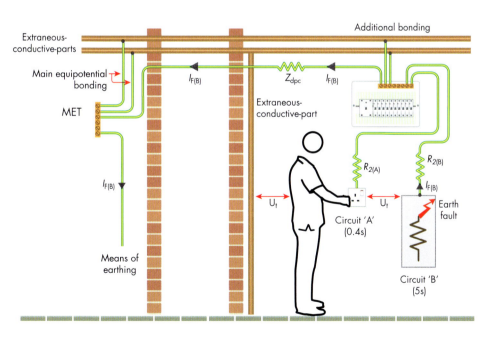

▶ **Figure 5.1** Example of touch voltages occurring in an installation under earth fault conditions

Whereas the purpose of earthing is to limit the duration of the touch voltages, the purpose of main equipotential bonding is to minimise the magnitude of these voltages. In essence, main equipotential bonding will limit touch voltages to acceptable levels until the earth fault is automatically disconnected.

There are two basic protective measures associated with main equipotential bonding:

▸ EEBADS: This measure is extensively used in electrical installations in the United Kingdom and elsewhere. Figure 5.2 illustrates an installation where there are main equipotential bonding conductors, and in a location where there is an increased risk of electric shock, supplementary bonding conductors.
▸ Earth-free local equipotential bonding: This is another protective measure against indirect contact. Unlike EEBADS, this measure is required to be under effective supervision. This seldom-used application is restricted to laboratories, electronic workshops and the like where an earth-free environment is required.

Supplementary equipotential bonding, sometimes referred to as additional bonding, is also employed as an additional protective measure to EEBADS. It is required by BS 7671 in certain of the special installations or locations addressed in Part 6 of BS 7671. Figure 5.2 illustrates examples of supplementary equipotential bonding conductors.

▸ **Figure 5.2** Examples of main and supplementary equipotential bonding conductors

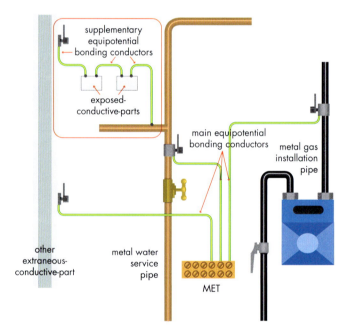

EEBADS is the most commonly used measure for protection against electric shock. A prerequisite for this measure to be effective is that the installation is required to be earthed, as well as having main equipotential bonding. During an earth fault, earthing permits an earth fault current to flow and to be detected by a device provided to automatically disconnect the circuit (e.g. an overcurrent protective device or an RCD). The provision of earthing plus a suitable protective device limits the duration of an earth fault to acceptable limits. On the other hand, main equipotential bonding minimises the touch voltage that may appear between exposed-conductive-parts and other exposed-conductive-parts as well as extraneous-conductive-parts, thereby together preventing the occurrence of dangerous voltages.

Although EEBADS is the most widely used technique, it should be noted that this protective measure does not obviate occurrences of indirect contact, nor does it prevent touch voltages being present during an earth fault.

The protective measure of EEBADS embraces three separately identifiable components:

▸ earthing of installation equipment metalwork (exposed-conductive-parts)
▸ equipotential bonding, as required, of non-electrical metalwork (extraneous-conductive-parts)
▸ automatic disconnection of supply.

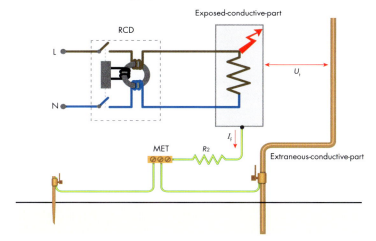

▸ **Figure 5.3** Illustration of main equipotential bonding

Figure 5.3 illustrates part of an electrical installation. The type of system of which it forms a part is not important here. In the circuit shown, an earth fault has developed in the current-using equipment and as a result a fault current (I_f) flows along the circuit protective conductor and back to the source. A small proportion of the current may flow through the main equipotential bonding conductor directly to Earth, and thence back to the source.

The touch voltage (U_t), or the potential difference, between the exposed-conductive-parts of the equipment and the simultaneously accessible extraneous-conductive-part is given by Equation (5.2):

$$U_t = I_f R_2 = \frac{U_0}{Z_s} \times R_2 \quad (V) \tag{5.2}$$

Equation (5.2) ignores the reactance of the circuit protective conductor, and any small effect due to currents flowing in the main equipotential bonding conductor.

To apply Equation (5.2), take the example of a circuit supplying an item of current-using equipment rated at 230 V (U_0) and 5 kW (resistive load). The circuit is protected by a 25 A, BS 88 fuse, and the earth fault loop impedance Z_s is 2.4 Ω. (the limiting value from Table 41D of BS 7671). R_2 is given as 1.1 Ω. The installation forms part of a TN-C-S system where the external earth fault loop impedance, Z_e, is 0.35 Ω.

The calculated touch voltage, U_t, is given in (5.3):

$$U_t = I_f R_2 = \frac{U_0}{Z_s} \times R_2 = \frac{230}{2.4} \times 1.1 = 105.4 \quad V \tag{5.3}$$

By connecting the MET to the extraneous-conductive-parts, the touch voltage, U_t, is minimised. Without this conductor, the potential difference would approximate to the voltage drop produced by the earth fault current, I_f, along the full length of the earth return path, and this could be significantly greater than $I_f R_2$ because of the voltage dropped across the supply protective conductor. Failure to provide all necessary main equipotential bonding conductors within an installation will almost certainly increase the risk of electric shock associated with indirect contact.

In the case cited in Equation (5.3), and assuming that the impedance of the supply protective conductor (the combined neutral protective conductor of the supply) is 0.15 Ω, the voltage dropped across the supply protective conductor would be approximately 14 V (0.15 x U_0/Z_s). Had not the main equipotential bonding been present, the touch voltage would have been approximately 120 V. With the main bonding present, the voltage is 14 V.

The EEBADS protective measure is achieved by co-ordinating the characteristics of the protective device for automatic disconnection and the relevant impedance of the circuit concerned, but the requirements for main equipotential bonding conductors and, where applicable, supplementary equipotential bonding are required to be met. Automatic disconnection, a necessary constituent of EEBADS, is addressed in Chapter 7 of this Guidance Note.

5.2 Main equipotential bonding conductors

Main bonding or more correctly main equipotential bonding is required in most electrical installations. Such bonding is an essential part of the most commonly used method of protection against indirect contact, namely EEBADS. Such bonding is also required where the method of protection against indirect contact is by earth-free equipotential bonding.

The first part of Regulation 413-02-02 requires that 'in each installation main equipotential bonding conductors complying with Section 547 shall connect to the main earthing terminal extraneous-conductive-parts of that installation including the following:

- water services pipes
- gas installation pipes
- other service pipes and ducting
- central heating and air conditioning systems
- exposed metallic structural parts of the building
- the lightning protection system'.

Regulation 413-02-02 goes on to require that 'where an installation serves more than one building the above requirements shall be applied to each building'.

The final paragraph of the Regulation stipulates that 'to comply with the Regulations it is also necessary to apply equipotential bonding to any metallic sheath of a telecommunication cable. However, the consent of the owner or operator of the cable shall be obtained'.

This Regulation requires all extraneous-conductive-parts to be bonded to the MET. An extraneous-conductive-part is defined in Part 2 of BS 7671 as:

> *A conductive part liable to introduce a potential, generally earth potential, and not forming part of the electrical installation.*

Section 8 dealing with extraneous-conductive-parts addresses this definition in more detail.

It should be noted that, whilst a number of examples of extraneous-conductive-parts are given in the Regulation, the list is not exhaustive and that other items not listed may well fall within the definition.

Extraneous-conductive-parts may be, and often are, connected to the MET individually, as shown in Figure 5.4. However, it is permitted to connect them collectively or in groups where the main equipotential bonding conductor is looped from one extraneous-conductive-part to another. Where bonding is undertaken in this way, the main equipotential bonding conductor should remain unbroken at intermediate points, as shown in Figure 5.5, thus maintaining continuity to other extraneous-conductive-parts should one be disconnected for whatever reason.

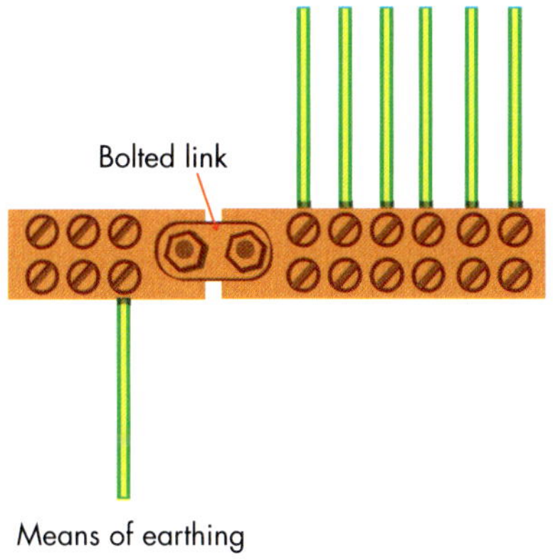

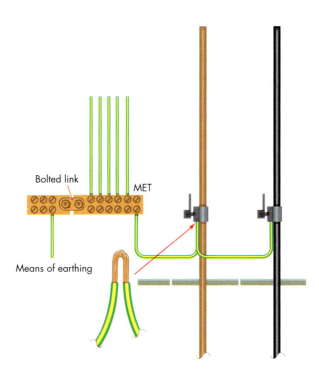

▶ **Figure 5.4** The MET

▶ **Figure 5.5** An unbroken main equipotential bonding conductor

Main equipotential bonding

Figure 5.6 shows typical main equipotential bonding conductors connecting extraneous-conductive-parts separately to the MET and hence to the means of earthing.

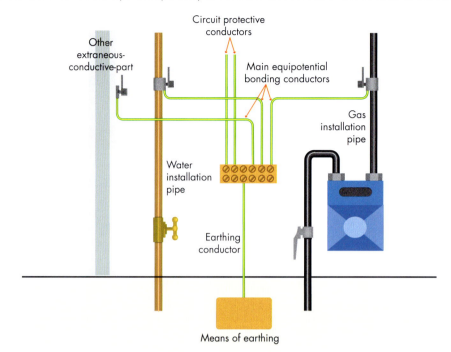

Figure 5.6 A typical example of main equipotential bonding

There are a number of types of conductor that are suitable for use as a main equipotential bonding conductor although a single-core non-flexible copper cable with a green-and-yellow covering is probably the most commonly used.

Regulation 543-02-02 also permits the use of other types of conductor, including metal parts of wiring systems such as metal conduits and metal trunking as well as metal sheaths or armouring of cables, provided that all the relevant requirements of BS 7671 are met. However, neither a gas nor an oil pipe may be used for such a purpose (see Regulation 543-02-01). Similarly, flexible or pliable conduit is precluded for such use by this Regulation.

For installations forming part of a TN-C-S system in which the supply is earthed at multiple points (PME conditions applying), and it is intended to utilise the armouring of a cable or a core of an armoured cable as a main bonding conductor the designer needs to consider the effects of the currents that may flow in the conductor due to network conditions. The armouring or a core of such a cable is normally not used as a main equipotential bonding conductor where PME conditions apply unless the electrical installation designer determines that the heat produced in the armouring or core due to its use will not cause overheating of the live conductors of the cable when on full load (Figure 5.7).

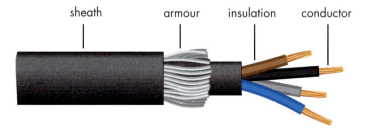

Figure 5.7 The armour is not recommended for use as a main equipotential bonding conductor in PME-supplied installations

5.2.1 Cross-sectional areas

Regulation 547-02-01 sets out the requirements for the minimum permitted CSA of a main equipotential bonding conductor. Where such a conductor is buried in the ground, the additional requirements of Regulation 542-03-01 for earthing conductors have to be applied, as detailed in Chapter 3.

For installations fed by a non-PME supply, Regulation 547-02-01 requires main equipotential bonding conductors to have a CSA of:

- not less than half that required for the earthing conductor, and
- not less than 6 mm², and
- need not be more than 25 mm² if the conductor is of copper or, if of another metal, a CSA affording equivalent conductance.

This regulation stipulates that the CSA of the main equipotential bonding conductors should be related to the CSA of the earthing conductor. Where the CSA of the earthing conductor is selected using Table 54G of BS 7671, as is normally the case, then Table 5.1 provides data relating the main equipotential bonding conductor to the supply phase conductor.

▶ **Table 5.1** CSAs of earthing conductors and main equipotential bonding conductors (non-PME) supplied installations

Phase conductor[1] CSA (mm²)	Earthing conductor[1,2,3] CSA (mm²)	Main equipotential bonding conductor[1] CSA (mm²)	Comment
4	4	6	Main equipotential bonding conductor CSA must have a minimum value of 6 mm²
6	6	6	
10	10	6	
16	16	10	
25	16	10	
35	16	10	
50	25	16	
70	35	25	
95	50	25	
120	70	25	
150	95	25	Main equipotential bonding conductor CSA need not be more than 25 mm²
185	95	25	
240	120	25	
300	150	25	
400	240	25	

1 Assumes that the conductors are copper.
2 Other constraints may apply to the CSA of the earthing conductor (see Chapter 3).
3 Conductor may be made up of more than one cable, thus reducing the overall CSA, if desired.

For a non-copper main equipotential bonding conductor, the 'equivalent conductance' requirement will be met if the CSA (S_m) of the conductor is not less than that given by Equation (5.4):

$$S_m \geq S_c \frac{\rho_m}{\rho_c} \text{ (mm}^2\text{)} \qquad (5.4)$$

where: S_m is the minimum CSA required for the main bonding conductor (in a metal other than copper)

S_c is the minimum CSA required for a copper main bonding conductor
ρ_m is the resistivity of the metal from which the main bonding conductor is made
ρ_c is the resistivity of copper.

For a steel main equipotential bonding conductor, the ratio ρ_m/ρ_c is approximately eight. Similarly, for an aluminium main equipotential bonding conductor, the ratio is 1.626. So for conductors made of these metals, the CSAs would be required to be eight and 1.626 times that of copper, respectively, in order to afford an equivalent conductance.

For installations supplied by a PME supply, Regulation 547-02-01 requires the main bonding conductors to be selected in accordance with the neutral conductor of the supply and Table 54H.

Because the earthing conductor also performs the function of a main equipotential bonding conductor, the requirements of Regulation 547-02-01 should also be met for the earthing conductor, so that the CSA of the earthing conductor is not less than that required for main equipotential bonding conductors, as well as meeting the requirements of Regulation 542-03-01.

Table 54H of BS 7671 sets out the requirements for the minimum CSAs of main equipotential bonding conductors. This data is available in Table 3.2 of this Guidance Note.

Where PME conditions apply, the electricity distributor may have particular requirements for the CSA of the main equipotential bonding conductor, which may exceed the minimum CSAs given in BS 7671. If there is doubt in this respect, the electricity distributor should be consulted and guidance sought at an early stage of the design.

The supply neutral conductor referred to in column 1 of Table 3.2 is the neutral conductor of the electricity distributor's low voltage supply to the installation. This is the combined protective and neutral PEN or CNE conductor of the supply. It is not the neutral conductor on the consumer's side of the supply terminals which may have a smaller CSA, as might be the case in a downstream distribution circuit feeding a separate building.

Where the use of non-copper main equipotential bonding conductors is contemplated, the advice of the electricity distributor should be sought.

5.2.2 Identification

For a main equipotential bonding conductor consisting of a single-core cable or a core of a cable, Regulation Groups 514-03 and 514-04 require it to be identified, wherever necessary, with the green-and-yellow colour combination, as shown in Figure 5.8. However, a bare main equipotential bonding conductor, such as a tape, strip or bare stranded conductor, is required by Regulation 514-04-06 to be identified, again where necessary, at intervals with the green-and-yellow colour combination.

As with all wiring, main equipotential bonding conductors have to be identifiable for inspection, testing, repair or alteration of the installation (see Regulation 514-01-02). To fulfil this requirement, every main equipotential bonding conductor is required to be marked or labelled to indicate its function, and to identify the item(s) to which it connects (e.g. structural steel or water service pipe) unless this is clear from the arrangement of the conductor, as might be the case from its position or from the terminals to which it connects, as illustrated in Figure 5.8.

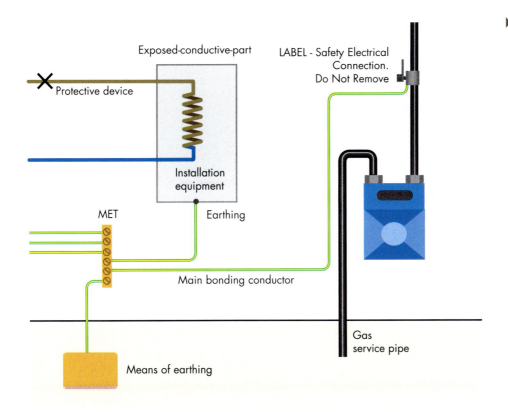

▶ **Figure 5.8**
Identification of protective conductors by the green-and-yellow colour scheme and labels

5.2.3 Supports

Where a main equipotential bonding conductor forms part of a composite cable, such as a separate core or the metal armouring or sheath, the method of support for the bonding conductor will be dictated by the type of cable and the manufacturer's installation instructions.

However, single-core main equipotential bonding conductors, and similar tape conductors, are required to be adequately supported, avoiding non-electrical services such as pipework as a means of support, so that they are able to withstand mechanical damage etc. (see Section 522 of BS 7671), and any anticipated factors likely to result in deterioration (see Regulation 543-03-01).

Table 5.2 provides guidance on providing adequate support for main equipotential bonding conductors and supplementary equipotential bonding conductors. Figure 7.9 illustrates main equipotential bonding conductors supported in an acceptable manner.

Overall cable diameter, ϕ (mm)	Horizontal spacing (mm)	Vertical spacing (mm)	Comment
<9	250	400	This data is provided as a guide and may be overridden by the constraints of good workmanship and by visual considerations
$9 < \phi \leq 15$	300	400	
$15 < \phi \leq 20$	350	450	
$20 < \phi \leq 40$	400	550	

▶ **Table 5.2**
Recommended spacing of supports of single core rigid copper cables for main equipotential bonding conductors

▶ **Figure 5.9** Main equipotential bonding conductor supports

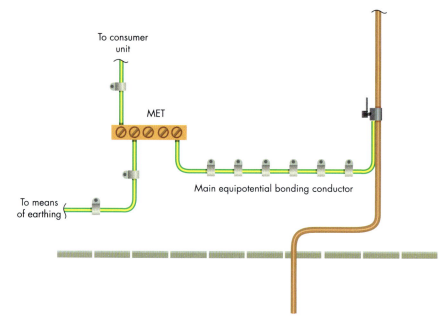

5.2.4 Alterations and extensions

Where alterations and/or extensions to an existing installation which pre-dates the current requirements of BS 7671 with regard to main equipotential bonding are to be performed, it is important to fully consider any shortcomings. For example, it may be that not all extraneous-conductive-parts are main equipotential bonded or it may be that the CSAs of the conductors are insufficient to meet the current requirements.

Where there is an absence of main equipotential bonding conductors to certain extraneous-conductive-parts, the obvious solution is to remedy this by installing suitable bonding conductors. However, where the deficiency is that the CSA of the conductor is less than currently required, the designer needs to consider carefully whether or not retaining the existing bonding conductors will be able to provide adequate safety levels and if not then should the conductors be upgraded to meet current requirements.

Where modifications to an existing installation are planned, the starting point should be the requirement of Regulation 130-07-01, reproduced here for convenience:

> *No addition or alteration, temporary or permanent, shall be made to an existing installation, unless it has been ascertained that the rating and the condition of any existing equipment, including that of the distributor, which will have to carry any additional load is adequate for the altered circumstances and the earthing and bonding arrangements are also adequate.*

It is clear that it is essential for the main equipotential bonding to be adequate and where such bonding is not present then it should be provided. It would be totally unacceptable to omit main equipotential bonding conductors or to rely on bonding conductors with inadequate CSAs.

For the case where all main equipotential bonding conductors are present but their CSAs are less than those currently required, the installation designer should make an evaluation of their suitability for use under the proposed new conditions.

Having considered the adequacy of the existing main equipotential bonding conductors, an installation designer may, in those cases in which the resulting degree of safety is not less than that obtained by compliance with BS 7671, elect to retain the existing conductors. However, the departure is required to be recorded on the certification for the work, both in the departures and in the comment on the existing installation (see Regulations 120-01-03 and 743-01-02).

In cases of installations fed by PME supplies, a careful assessment should be made before relying on bonding conductors with inadequate CSAs. For such installations, Table 54H of BS 7671 sets out the minimum CSAs which are based on those given in *Electricity Supply Regulations 1988* (now superseded by the *ESQCR*). Prior to the publication of these statutory regulations, smaller minimum CSAs were accepted by some electricity boards. However, the minimum CSAs given in Table 54H of BS 7671 were intended to obviate the likelihood of main bonding conductors overheating due to the PME network circulating currents. Only where the designer is comfortable with the adequacy of the existing bonding conductors should the decision be made not to replace them with ones meeting the current minimum requirements.

5.3 Earth-free equipotential bonding

Earth-free equipotential bonding is one of the five protective measures against indirect contact permitted by BS 7671.

This protective measure is intended to prevent the appearance of a dangerous touch voltage between simultaneously accessible conductive parts in the event of a failure of the basic insulation. Such a protective measure should only be used where the installation is being supervised by a suitably qualified electrical engineer.

Local equipotential bonding is required to be carried out by connecting together all the exposed-conductive-parts and extraneous-conductive-parts within the location of the installation. It is important that this bonding is not connected to Earth. This way, persons within the location are free from electric shock. However, where a person enters or leaves the equipotential location there may be a risk of electric shock, and BS 7671 requires precautions to be taken to address this risk. This includes a requirement to place a notice in a prominent position adjacent to every point of access to the location, as illustrated in Figure 5.10.

This measure is strictly limited to situations that need to be earth-free. Examples include medical, special electronic and communications equipment applications, in which the equipment, while having exposed-conductive-parts (i.e. not being Class II equipment or having equivalent insulation), will not operate satisfactorily where connected to a means of earthing. The locations where the measure is applied may have a non-conducting floor, or a conductive floor which is insulated from earth and to which every accessible exposed-conductive-part within the location is connected by local equipotential bonding conductors. The location requires effective supervision and regular inspection and testing of the protective features.

The correct implementation of the requirements results in a 'Faraday cage' and prevents the appearance of any dangerous voltages between simultaneously accessible parts within the location concerned. Thus, in this application of equipotential bonding, the use of the word 'equipotential' really does reflect what it means for a first fault on the equipment, although this may not hold true for two simultaneous faults.

This measure cannot be realistically applied to an entire building and it is difficult to co-ordinate safely with other protective measures used elsewhere in the installation. In particular, precautions are necessary at the threshold of the earth-free equipotential location. Also, as mentioned earlier, a warning notice is required to be fixed at every point of access into the location to warn against the importation of an earth.

The form of the supply to the equipment requires special consideration. Using the mains supply of a TN system would import an earth into the location via the earthed neutral conductor which, in the event of a neutral fault to an exposed-conductive-part, would earth the equipment.

Earth-free local equipotential bonding is normally associated with electrical separation which overcomes this problem, but where two measures of protection are to be used in the same location care should be taken to ensure that the particular requirements for each measure are fully satisfied and, most important, are mutually compatible.

Not surprisingly, this protective measure is not suitable for locations where there is perceived to be an increased risk of electric shock.

Regulation Groups 413-05 and 471-11 and Regulation 514-13-02 refer.

▶ **Figure 5.10** Label for earth-free location

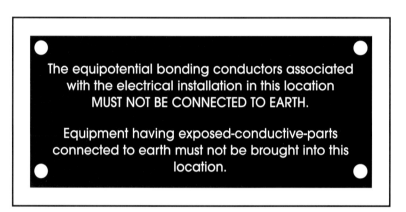

5.4 Bonding of lightning protection systems

Although BS 7671 excludes lightning protection systems of buildings from its scope, there is a requirement, in Regulation 413-02-02, to provide a main equipotential bonding connection of such systems to the MET of the electrical installation. The requirements in terms of CSA of this bonding are as they are for all other extraneous-conductive-parts. The recommendations of BS 6651 are also required to be taken into account.

Based on the guidance given in BS 6651, the positioning of main bonding connections to a lightning protection system is important and best determined by a lightning protection system designer.

Where the down conductors of the lightning protection system are systematically connected together, a single main bonding connection to that system will suffice. A connection from the MET would normally be taken to the closest down conductor by the most direct route.

Although each bonding connection arrangement should be considered on its own merits in consultation with all the authorities concerned, Figure 30 of BS 6651: 1999 is reproduced here as Figure 5.11 and shows a typical arrangement for such a connection.

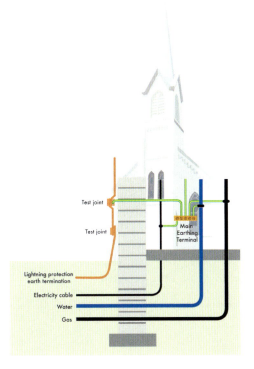

▶ **Figure 5.11** Typical arrangement for the main bonding conductor connecting to a lightning protection system

Additionally, it may be necessary to connect parts of the electrical installation to adjacent parts of the lightning protection system which are separated by less than the predicted isolation distance necessary to prevent side-flashing.

Bonding connections to a lightning protection system are often made outdoors and, where this is the case, attention should be paid to any external influences that might affect the connection. This is particularly important where conductors with different shapes and made from different metals are used, and precautions may need to be taken to avoid corrosion and to create joints with an adequate mechanical strength and durable electrical continuity.

In exceptional cases, the designer of a lightning protection system may decide, for reasons of safety, that a connection to the MET should not be made. For the electrical installation designer, this would be a departure from Regulation 413-02-02 and as such this is required to be recorded on the Electrical Installation Certificate and drawn to the customer's attention.

5.5 Extraneous-conductive-parts common to a number of buildings

Industrial and commercial facilities sometimes consist of a number of separate buildings supplied from a common electricity source, or separately fed with their own supplies. It is often a feature of these premises that they have extraneous-conductive-parts that are common to a number of separate buildings. For example, a common metal pipework system transporting oil or gases between buildings requires careful consideration. Figure 5.12 illustrates three buildings that have a common pipework system.

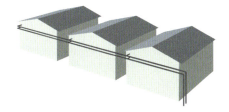

▶ **Figure 5.12** Three buildings with a common pipework system

In these circumstances, the question of main equipotential bonding of the extraneous-conductive-parts needs careful thought. The guidance given here relates to installations fed from a single low voltage source (a single low voltage distribution transformer). Two situations are considered:

▶ Situation 1: The three buildings are supplied from a common source with the electricity distributor's supply to one of the buildings, and distribution circuits emanating to the other two buildings (Figure 5.13).
▶ Situation 2: The three buildings are supplied from a common source with the electricity distributor's separate electricity supply to each of the three buildings (Figure 5.14).

▼ **Figure 5.13** Three buildings with a single electricity supply

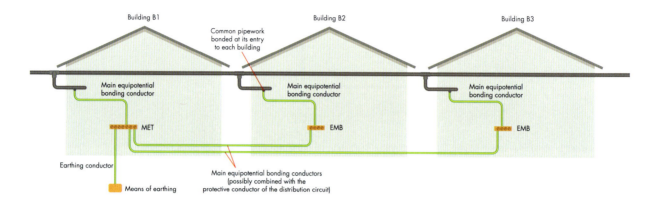

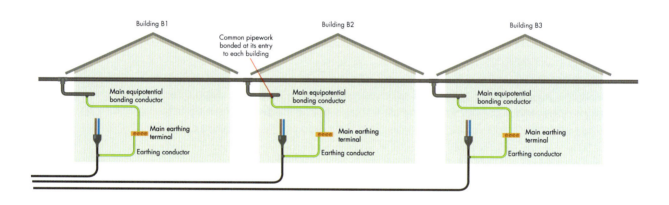

▲ **Figure 5.14** Three buildings with three separate electricity supplies

Where multiple sources, including sources from local transformers, are involved, further advice should be sought from a suitably qualified electrical engineer.

A metal pipework system that enters a building would be considered to be an extraneous-conductive-part. This would be so even where the pipework has previously exited another building in which it has been separately bonded elsewhere. The pipework is then likely to introduce a potential, either the earth potential or earth fault potential from the building into which the pipework previously entered, and hence it would fall within the definition of an extraneous-conductive-part.

There would therefore be a requirement to provide main equipotential bonding to the common pipework system in each building that it enters, either to the MET if the origin of the supply is there, or to the EMB (earth marshalling bar) if in another building.

In Figures 5.13 and 5.14, each building will have its own earth marshalling point, either in the form of a MET or an EMB with the common pipework system main equipotential bonded to the MET or EMB in each building:

In situation 1, the MET is in building B1 and there are EMBs in buildings B2 and B3. The common pipework system is main bonded:

▸ to the MET in building B1
▸ to the EMB in buildings B2 and B3.

In situation 2, a MET exists in buildings B1, B2 and B3. The common pipework system is main bonded:

▸ to the MET in building B1
▸ to the MET in building B2
▸ to the MET in building B3.

The position of the connection of the main equipotential bonding conductors to the common pipework system should be as close as practicable to the point of entry of the pipework into each building.

The main equipotential bonding conductors between the MET and the two EMBs may be made utilising the circuit protective conductors of the distribution circuits to buildings B2 and B3, provided that the circuit protective conductors meet the CSA requirements for both functions.

5.6 Installations serving more than one building

Where a supply serves installations in more than one building, Regulation 413-02-02 requires that main equipotential bonding conductors should be applied in each building. Therefore, the extraneous-conductive-parts in all of the buildings should be connected to the MET of which there is only one.

So, where a supply serves three separate buildings, designated B1, B2 and B3, each will be required to be provided with main equipotential bonding conductors to all extraneous-conductive-parts, complying with Section 547 of BS 7671.

In such a scenario, a MET is located in building, B1 at a position adjacent to the incoming supply at the origin of the installation. The two other buildings each have EMBs. An EMB marshals all the main equipotential bonding conductors for a particular building and also provides for the connection of a conductor which runs to the MET.

As with any main equipotential bonding, the CSA of the conductors will depend on the system type. Figure 5.15 illustrates a three-building complex where the system type is TN-S for buildings B1 and B3, and TT for building B2. Figure 5.16 illustrates a three-building complex where the system type is TN-C-S with PME for buildings B1 and B3, and TT for building B2.

For buildings B1 and B3, it is important to note that irrespective of the system type the CSA of the main equipotential bonding conductors should not be less than that given in Regulation 547-02-01. This applies equally to the conductors in both buildings in the examples given in Figures 5.15 and 5.16. For the three system types cited, the requirements in terms of CSA of the main equipotential bonding conductors are:

- For TN-S and TT systems: The CSA of the bonding conductor is required to be not less than half the CSA required for the earthing conductor, subject to a minimum of 6 mm^2, but it need not exceed 25 mm^2 if the conductor is made of copper (or affording an equivalent conductance in other metals). Note that the earthing conductor, of which there is but one, is the conductor that connects the MET to the means of earthing, which in this case is the electricity distributor's earthing terminal.
- For TN-C-S systems: Where PME conditions apply, the main bonding conductor CSA has to be selected in relation to the supply neutral conductor and in accordance with Table 54 H of BS 7671. Note that the supply neutral conductor is the neutral conductor upstream of the electricity distributor's cut-out.

Both figures illustrate three separate buildings, each with a single extraneous-conductive-part in the form of an incoming metal service pipe. In reality, other extraneous-conductive-parts, such as structural steel and/or lightning protection systems, may be present but for simplicity and clarity in the figures only one such item is shown.

The application of Regulation 547-02-01 for complexes with a number of separate buildings can result in the CSA of the main equipotential bonding conductors being required to be greater than the CSA of the associated live conductors of the distribution circuit supplying that building, particularly where PME conditions apply. This is illustrated in Figure 5.16 where the main equipotential bonding conductor in building B3 is required to have a CSA of 35 mm^2, since the CSA of the supply neutral is 120 mm^2, which is considerably greater than the required CSA of 16 mm^2 for the circuit protective conductor of the distribution circuit to building B3.

Where a circuit protective conductor also acts as a bonding conductor the requirements for both functions will have to be met. In other words, as well as meeting the requirements of Regulation Group 543-01 for a circuit protective conductor (CPC), the requirements of Regulation 547-02-01 also have to be met. This protective conductor dual function has been adopted for the distribution circuit to building B2 in Figures 5.15 and 5.16.

For the case of building B2 where the distribution circuit includes a CPC, as might be the case with an armoured cable, it is important to consider Regulation 542-01-09, which reads:

> *Where a number of installations have separate earthing arrangements, any protective conductors common to any of these installations shall either be capable of carrying the maximum fault current likely to flow through them or be earthed within one installation only and insulated from the earthing arrangements of the other installation. In the latter circumstances, if the protective conductor forms part of a cable, the protective conductor shall be earthed only in the installation containing the associated protective device.*

In the examples shown in Figures 5.15 and 5.16, the distribution CPC (cable armouring) has been insulated from the earthing arrangement of building B2 by means of an insulated cable gland. This effectively provides for the two earthing arrangements to be considered simultaneously inaccessible, thereby obviating the need to apply Regulation 413-02-03 which requires the CPC and the earthing arrangement to be connected together where they are simultaneously accessible, in order to prevent the associated risk of an electric shock due to the potential difference between the two points. However, the armouring would be required to be earthed in building B1.

Where a protective conductor connection is to be provided between the earthing arrangement of building B2 and that of buildings B1 and B3, then the protective conductor common to the two installations would be required by Regulation 542-01-09 to be capable of carrying the maximum fault current likely to flow through it. As a consequence, the magnitude and duration of the fault current would have to be calculated and, by using Equation (3.1), the minimum CSA required for the conductor would have to be determined.

For an installation which is fed from a supply to which PME conditions apply, but which forms part of a TT system as in building B2, these conditions do not apply to that installation provided the PME earthing has not been used to earth the installation.

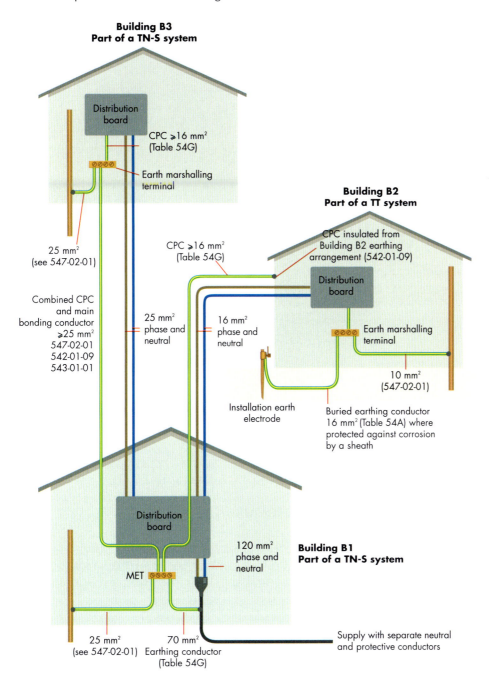

▶ **Figure 5.15** A three-building complex – system type is TN-S

Main equipotential bonding

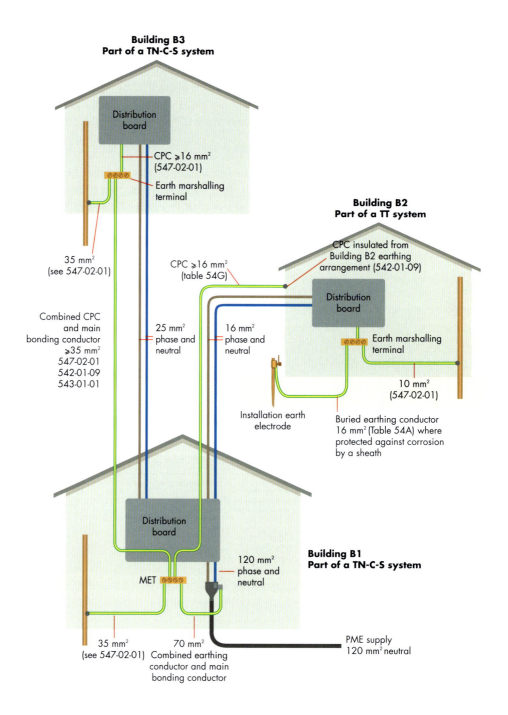

Figure 5.16 A three-building complex – system type is TN-C-S

5.7 Multi-occupancy premises

Main equipotential bonding in multi-occupancy premises can sometimes present problems in terms of interpreting the requirements of BS 7671. This typically involves premises such as blocks of flats, office blocks and shops that are separately let and have separate electricity supplies. Additionally, single-occupancy properties that have subsequently been converted into separate units sometimes also present problems.

BS 7671 requires main equipotential bonding conductors in each installation. The definition of installation, reproduced below together with that for the origin of the installation, relates to an installation supplied from a common origin.

> **Electrical installation.** (abbr.: Installation). An assembly of associated electrical equipment supplied from a common origin to fulfil a specific purpose and having certain co-ordinated characteristics.

Origin of an installation. *The position at which electrical energy is delivered to an electrical installation.*

It is clear that BS 7671 requires main equipotential bonding in each and every installation connecting together extraneous-conductive-parts to the MET and this would apply equally to separate installations on a multi-occupancy building.

The CSA of every main equipotential bonding conductor has to be in accordance with Regulation 547-02-01 and will depend on the system type as discussed in Clause 5.2.1. As previously mentioned, except where PME conditions apply, the CSA of the main bonding conductor is related to the CSA of the earthing conductor of the installation (i.e. in each separate installation). For PME-supplied installations, the CSA of the main equipotential bonding conductors is related to the electricity distributor's supply neutral conductor (not the neutral conductor downstream of the cut-out on the consumer's side, which may have a different CSA).

Figure 5.17 illustrates a typical main bonding arrangement in a small office block. Each office installation forms part of a TN-C-S system with a separate PME supply.

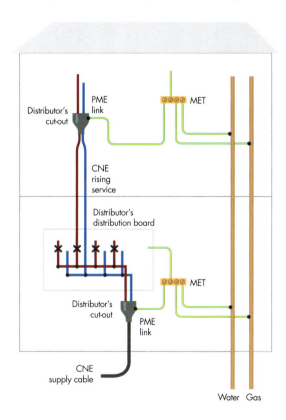

▶ **Figure 5.17** Typical main bonding arrangement in a small office block

As with all electrical installations, the electricity distributor may have additional requirements for main bonding, especially in the case where PME conditions apply. It is important therefore that the installation designer seeks guidance in this respect from the electricity distributor.

The connection of main equipotential bonding conductors to extraneous-conductive-parts, such as gas and water service pipes, should be made in accessible locations as near as practicable to the point of entry into the office building.

6 Extraneous-conductive-parts and their connections

6.1 Definition of an extraneous-conductive-part

It is important to note that the term extraneous-conductive-part is hyphenated and it is therefore a single term. This means it has a special meaning which could not be attributed had it not been so hyphenated.

The term is used extensively with regard to equipotential bonding (protection against indirect contact), but it is also used in connection with other measures for protection against electric shock such as SELV (Separated Extra-Low Voltage), placing out of reach and non-conducting locations.

An extraneous-conductive-part is defined in BS 7671 as:

> *A conductive part liable to introduce a potential, generally earth potential, and not forming part of the electrical installation.*

Even with such a definition, it is sometimes difficult to establish what is and what is not an extraneous-conductive-part. To assist in making this decision, the installation designer has to analyse the definition by breaking it down into three discrete parts:

- a conductive part
- liable to introduce a potential, generally earth potential and
- not forming part of the electrical installation.

Generally, it is metals with their high conductivity values that are described as the conductive part in the context of considering an extraneous-conductive-part. The conductivity of a material is a measure of its ability to conduct electricity or, put another way, of its conductance G, which is the reciprocal of resistance (as conductivity is the reciprocal of resistivity).

However, although non-metals are generally non-conductive, pipework systems which are intended to carry fluids may not be so described. Water in its purest form is a good insulating material with a 10^2 to 10^5 Ωm volume resistivity, which is almost as good an insulating material as, say, flexible PVC. Raw water which might be expected from the main water supply and water used in the circulation system of central heating systems that contains additives may well be laden with sufficient impurities to make the water conductive to some extent.

The potential referred to in the definition is generally taken to be that of the conductive mass of Earth. This is by convention taken as 0 V.

Metal pipework and other extraneous-conductive-parts, such as steel stanchions, entering the equipotential zone are generally considered to be at 0 V. However, it is not just extraneous-conductive-parts that are earthy that the designer needs to consider. A potential related to the earth fault in another building may be imported along an extraneous-conductive-part such as pipework.

Even when an extraneous-conductive-part is raised to a potential it may not be capable of transferring its potential to a person (or livestock) if it cannot be touched by a person in simultaneous contact with another conductive part, such as an exposed-conductive-part or another extraneous-conductive-part at a different potential, or a live part.

Therefore, for a conductive part to be able to transfer a potential to a person (or livestock) in contact with the potential of any of the conductive parts, that part should be accessible to be touched by such a person (or livestock).

On occasions, a conductive part may be partially insulated from Earth potential by virtue of some insulating material intervening between the conductive part and earth. This might be the case, for example, where an insulating board used as part of the building construction places a high resistance between the two points. In such cases, Equation (6.1) should be used:

$$R_{CP} > \left(\frac{U_0}{I_B}\right) - Z_{TL} \qquad (6.1)$$

where: R_{CP} is the measured resistance between the conductive part concerned and the MET of the installation (in ohms),
U_0 is the nominal voltage to Earth of the installation (in volts),
I_B is the value of current through the human body (or livestock) which should not be exceeded (in amperes), and
Z_{TL} is the impedance of the human body or livestock (in ohms).

Where the measured resistance R_{CP} satisfies the value calculated using Equation (6.1), the conductive part should not be considered to be an extraneous-conductive-part. However, the designer's decision should also take into account the likely stability of the resistance of the conductive part over the lifetime of the installation.

British Standard Published Document PD 6519, IEC 60479 *Guide to effects of current on human beings and livestock* provides data for Z_{TL} and I_B. For hand-to-hand contact, the value of the body impedance is given as 1000 Ω in dry conditions where U_0 is 230 V. The designer can then select a value of I_B between the two extremes:

▸ 0.5 mA – the threshold of perception
▸ 10 mA – the let-go threshold
▸ Using Equation (6.1) to illustrate the range of values of resistance R_{CP} for currents from 0.5 to 10 mA (0.5 mA being the threshold of perception and 10 mA being the let-go threshold) we get Equation (6.2) and Equation (6.3), respectively:

$$R_{CP} > \left(\frac{U_0}{I_B}\right) - Z_{TL} = \left(\frac{230}{0.5 \times 10^{-3}}\right) - 1000 = 460\,000 - 1000 = 459 \text{ k}\Omega \qquad (6.2)$$

$$R_{CP} > \left(\frac{U_0}{I_B}\right) - Z_{TL} = \left(\frac{230}{10 \times 10^{-3}}\right) - 1000 = 23\,000 - 1000 = 22 \text{ k}\Omega \qquad (6.3)$$

From Equation (6.3) we can see that if the designer elects to accept the let-go threshold as a safe level, the resistance of the extraneous-conductive-part to the MET

(R_{CP}) is above the threshold of 22 kΩ, then the conductive part need not be considered to be an extraneous-conductive-part. On the other hand, the designer would need to consider possible variations in resistance and whether a lower limit on the current flowing through the human body or livestock is necessary.

The definition of extraneous-conductive-part does not include conductive parts that form part of the electrical installation in question.

6.2 Some examples of extraneous-conductive-parts

Regulation 413-02-02 of BS 7671 provides a list of parts which may meet the definition of an extraneous-conductive-part:

- water service pipes
- gas installation pipes
- other service pipes and ducting
- central heating and air conditioning systems
- exposed metallic structural parts of a building
- a lightning protection system.

This list should not be regarded as exhaustive. Figure 6.1 shows a typical layout of extraneous-conductive-parts and associated main equipotential bonding conductors.

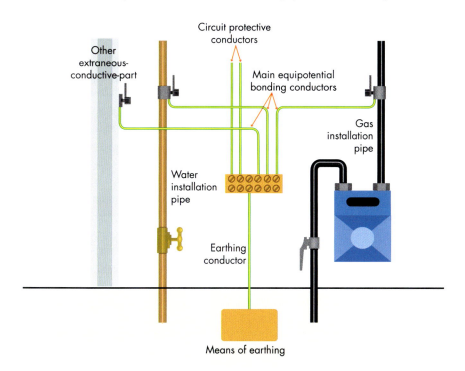

Figure 6.1 Typical main equipotential bonding

6.3 An example of a conductive part which is not an extraneous-conductive-part

It is necessary to look a little deeper when deciding about the bonding of metalwork accessible to people or livestock outside a building and in contact with the ground. For example, a metal window frame inserted in a brick building is not likely to introduce a potential and would not therefore present a hazard.

If, on the other hand, it has been main or supplementary bonded it may, under earth fault conditions, attain a considerable rise in potential during the permitted 5 s clearance of a fault. This could be dangerous to a window cleaner outside the building, with wet hands and standing on damp ground or on a metal ladder.

A metal window frame in a metal-clad building could, however, introduce an earth potential into the location. In such a case the metal-clad building would itself be an extraneous-conductive-part and would require main bonding, and there would be no requirement to repeat this bonding for the window. However, if the metal window is in a special location such as a bathroom, there may be a requirement to supplementary bond it to other extraneous-conductive-parts and exposed-conductive-parts.

6.4 Connection to pipework

BS 951: *1999 Electrical earthing – clamps for earthing and bonding – Specification* is the performance specification for clamps for connection of main and supplementary bonding conductors to pipes, solid rods and the like with circular cross-sections. The specification embraces the connection of:

- earthing conductors, having a CSA in the range from 2.5 to 70 mm^2, to earth electrode rods or other means of earthing
- bonding conductors to metal tubes of circular cross-section that have circumferences of not less than 18.8 mm (i.e. diameters of not less than 6 mm).

It is important to note that BS 951 clamps are not intended, and are therefore not suitable, for connection to the armouring or lead sheath of a cable. Clamps applied to an armoured cable can have the effect of crushing the bedding or insulation, thus reducing its effectiveness. Similarly, a clamp fitted to a lead cable sheath can cause the lead to cold-flow, and, with continuing expansion and contraction expected under varying load conditions, this can result in a high-resistance connection. This, in turn, would increase the earth fault loop impedance and may adversely affect disconnection times.

Nor is the clamp intended for any use other than to encircle a pipe or rod; for example, it should not be used to connect to an item that is too large for the clamp.

BS 951 lays down important mechanical constraints for metal clamps that are used for the purpose of connecting conductors to provide mechanically and electrically sound earthing and bonding connections to metal tubes.

Figure 6.2 shows a correctly installed BS 951 clamp connecting a copper protective conductor and a copper tube. The slots in the label are intended only as an aid to packaging and storage.

Figure 6.2 BS 951 clamp

BS 951 calls for the clamp to consist of:

1. a device for making electrical contact with the tube
2. a means of tightening the device on to the tube
3. a means of locking the arrangement given in item 2
4. a termination separate from the arrangement given in item 2 for attaching the protective conductor to the device given in item 1.

From the above, it should be noted that the clamp for the protective conductor is separate from the means of tightening and locking the clamp to make electrical contact with the tube.

A screw termination is capable of accepting one of the following:

▶ a conductor clamped under a screw head provided with a captive washer so that the screw head does not act directly on the conductor (also capable of accepting a looped unbroken conductor), or
▶ a single conductor clamped directly by a screw-threaded arrangement, the CSA of the conductor being within the range specified in Table 1 of BS 951 (also capable of accepting a looped unbroken conductor), or
▶ a bolted-on cable socket from a range of sockets that can accommodate conductors having CSAs covering the whole range specified in Table 1 of BS 951.

Table 6.1 replicates the data relating to termination reference and conductor CSAs given in Table 1 of BS 951.

▶ **Table 6.1** Termination reference and conductor size

Termination reference	Nominal CSA of conductor (mm^2)
A	2.5
B	4
C	6
D	10
E	16
F	25
G	35
H	50
I	70

Clamps are made of different metals in order to suit differing environments and the consequential level of corrosion. Regulation 522-05-01 of BS 7671 demands that such clamps are suitably protected from the effects of corrosion, or manufactured from material resistant to corrosion.

Figure 6.3 illustrates three clamps made from different metals and differing lengths, together with a warning notice required at each clamp, as required by Regulation 514-13-01, with the words 'SAFETY ELECTRICAL CONNECTION – DO NOT REMOVE'. BS 951 clamps are available in three colour types:

- red for dry, non-corrosive atmosphere
- blue for corrosive or humid conditions
- green for corrosive or humid conditions, and for larger sizes of conductor.

And three standard band sizes:

- to suit pipes with diameters of 12 – 32 mm
- to suit pipes with diameters of 32 – 50 mm
- to suit pipes with diameters of 50 – 75 mm.

Figure 6.3 Three clamps of different metals and differing lengths

Regulations 522-05-02 and 522-05-03 require that metals liable to initiate electrolytic action should not be placed in contact with one another. This would be relevant to clamps and their contact with a dissimilar metal and in particular to the aluminium warning notice label which is liable to severe corrosion where in contact with other metals.

With certain exceptions, all connections and joints are required to be accessible for inspection, testing and maintenance purposes (see Regulation 526-04-01). Earthing and bonding clamps are no exception to this requirement, and the clamp and its conductor termination are required to be accessible. Additionally, such clamps are required to be complete with their warning notice as shown in Figure 6.4.

Figure 6.4 A BS 951 earthing and bonding clamp warning label

BS 951 clamps may not be able to accommodate pipes and other circular components with large CSAs that are sometimes encountered in the construction and services of buildings. The answer is not to join two or more BS 951 clamps together, but to use a suitable proprietary clamp, as shown in Figure 6.5.

6.5 Connections to structural steel and buried steel grids

BS 951 clamps are suitable for use with pipes and rods with circular CSAs. However, extraneous-conductive-parts often exist in the form of irregular shapes that are required to be used as main equipotential bonding conductors and/or used in supplementary equipotential bonding. Suitable clamps are freely available from suppliers; however, the installation designer must only use clamps manufactured from materials that will not adversely affect the extraneous-conductive-part to which it connects. Table 6 of BS 7430 provides recommendations for the manufacture of earthing components.

As with all connections, these clamps are required to meet the requirements of Regulation 526-01-01, which requires every connection to provide durable electrical continuity and adequate mechanical strength. This Regulation also requires that the connection be suitable for the conductors. The clamp should therefore be suitable for connection to the extraneous-conductive-part, in terms of its shape, CSA and dimensions, as well as being able to act as a supplementary equipotential bonding conductor. The clamp should be equipped with a means of locking to prevent loosening due to, for example, vibration.

▶ **Figure 6.5** Clamps for larger diameter pipes etc.

Figure 6.6 illustrates some typical examples of clamps for connecting main equipotential bonding conductors to steel building components.

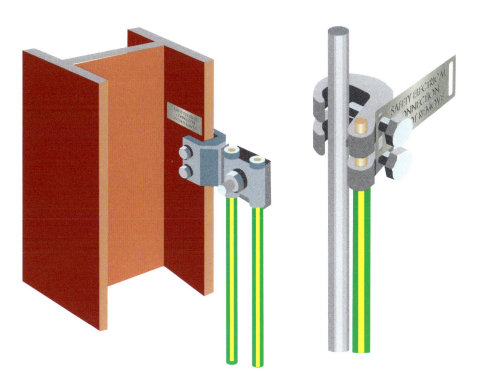

▶ **Figure 6.6** Typical examples of clamps for connecting to steel building components

Automatic disconnection 7

7.1 Automatic disconnection

EEBADS provides protection against indirect contact by virtue of co-ordinating the characteristics of the protective device for automatic disconnection and the relevant impedance of the circuit concerned. To achieve this, BS 7671 provides for three basic requirements:

- installation equipment metalwork (exposed-conductive-parts) is required to be connected to the earthing arrangement, and
- main equipotential bonding and, where applicable, supplementary equipotential bonding, is required for any item of non-electrical metalwork falling within the definition of an extraneous-conductive-part, and
- automatic disconnection of the supply under earth fault conditions is required.

Automatic disconnection is an indispensable ingredient of EEBADS which is by far the most common form of protection against indirect contact electric shock.

Whatever the type of system, the operating characteristic of each protective device for automatic disconnection and the earth fault loop impedance of the associated circuit should be properly co-ordinated so as to achieve the automatic disconnection within the prescribed time limit set out in BS 7671.

In order to meet the prescribed time limits for automatic disconnection, Regulation 413-02-04 requires coordination between:

- the time/current characteristic of each protective device for automatic disconnection,
- the earthing arrangements, and
- the relevant impedance of the circuit concerned.

This co-ordination is required to make certain that under earth fault conditions the potentials between simultaneously accessible exposed-conductive-parts and extraneous-conductive-parts are of such magnitude and duration as not to cause danger.

7.2 TN systems

For TN systems, automatic disconnection within a specified time is fulfilled in TN systems when Equation (7.1), given in Regulation 413-02-08, is satisfied:

$$Z_s \leq \frac{U_0}{I_a} \ (\Omega) \tag{7.1}$$

where: Z_s is the earth fault loop impedance in ohms,

I_a is the current causing the automatic operation of the disconnecting device within the time stated in Table 41A of BS 7671 as a function of the nominal voltage U_0 or, under the conditions stated in Regulations 413-02-12 and 413-02-13, within a time not exceeding 5s, and

U_0 is the nominal phase voltage to Earth.

For convenience, the data and information contained in Table 41A of BS 7671 is replicated here in Table 7.1.

▶ **Table 7.1** Table 41A of BS 7671: *Maximum disconnection times for TN systems* (see Regulation 413-02-09)

Installation nominal voltage U_0 (V)	Maximum disconnection time t (s)
120	0.8
230	0.4
277	0.4
400	0.2
greater than 400	0.1

Notes:
1 For voltages which are within the supply tolerance band (230 ± 10%), the disconnection time appropriate to the nominal voltage applies.
2 For intermediate values of voltage, the higher value of the voltage range in the table is to be used.
3 For certain special locations these maximum disconnection times are lower.

Table 7.1 sets out the prescribed time limits for automatic disconnection in TN systems for:

▶ all socket-outlets
▶ other final circuits supplying portable equipment intended for manual movement during use
▶ hand-held Class I equipment (Regulation 413-02-09)
▶ circuits supplying fixed equipment outside the earthed equipotential zone where the equipment has exposed-conductive-parts which may be touched by a person in contact directly with the general mass of Earth (Regulation 471-08-03).

The prescribed time limits for automatic disconnection in TN systems set out in Table 7.1 do *not* apply to:

▶ a final circuit supplying only stationary equipment
▶ an item of stationary equipment connected by a plug and socket-outlet where precautions are taken to prevent the use of the socket-outlet to supply hand-held equipment
▶ reduced low-voltage circuits (as described in Regulation Group 471-15)
▶ distribution circuits.

For those circuits and equipment for which Table 41A of BS 7671 does not apply, a disconnection time of not more than 5 s is permitted (Regulations 413-02-09 and 413-02-13 refer). A disconnection time of up to 5 s is only permitted where the exposed-conductive-parts of the equipment concerned and any extraneous-conductive-parts are within the earthed equipotential zone. However, Regulation 471-08-03 requires that disconnection occurs within the time stated in Table 41A of BS 7671 where a circuit supplies fixed equipment outside the earthed equipotential zone and the equipment has exposed-conductive-parts which may be touched by a person in contact directly with the general mass of Earth.

For some locations and installations where there is an increased risk of electric shock such as installations on construction sites and agricultural and horticultural premises (Sections 604 and 605 of BS 7671 respectively), the maximum disconnection times are reduced, as given in Table 7.2. Accordingly, the maximum earth fault loop impedance values are correspondingly lower.

Installation nominal voltage U_0 (V)	Maximum disconnection time t (s)
120	0.35
200 – 277	0.20
400, 480	0.05
580	0.02

▶ **Table 7.2** Tables 604A and 605A of BS 7671: *Maximum disconnection times for TN systems on construction sites and agricultural and horticultural premises* (see Regulations 604-04-02 and 605-05-02)

7.2.1 Earth fault loop impedance

For a circuit with a nominal voltage to Earth (U_0) of 230 V, the earth fault loop impedance values given in Tables 41B1, 41B2 and 41D of BS 7671 appropriate to the required disconnection time may be used for the types and ratings of the overcurrent devices listed. For some special installations where there is perceived to be an increased risk of electric shock, such as installations on construction sites and agricultural and horticultural premises (Sections 604 and 605 of BS 7671 respectively), the maximum disconnection times are reduced, and so too are the maximum earth fault loop impedances.

For convenience, Tables 7.3 to 7.6 give the limiting earth fault loop impedances for different disconnection times and for a number of common overcurrent protective devices. The tables also include 80 per cent 'rule of thumb' values for comparison with measured values of the earth fault loop impedance, Z_s. This is necessary where the measurements are taken at an ambient temperature of 20 °C and where the normal operating temperature for the conductors is 70 °C. Figure 7.1 shows the relationship between conductor resistance and temperature. Equation (7.2) allows an assessment to be made of the likely increase in resistance from that at an ambient temperature of 20 °C and normal operating temperature of 70 °C.

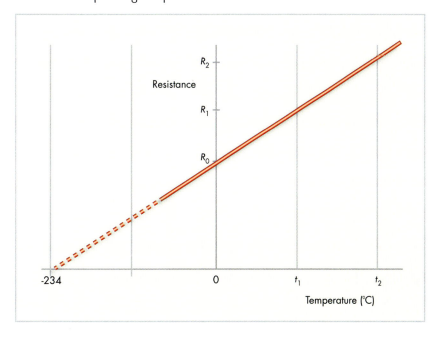

▶ **Figure 7.1** Conductor resistance against temperature

Equation (7.2) allows the resistance to be assessed at an elevated temperature if it is known at a lower temperature:

$$R_t = R_{20}\left[1+\alpha_{20}(t-20)\right] \ (\Omega) \tag{7.2}$$

where: R_{20} is the conductor resistance at 20 °C
R_t is the conductor resistance at temperature t
α_{20} is the resistance/temperature coefficient at 20 °C

To establish the resistance of a copper conductor of 2 Ω resistance at a temperature at 20 °C when at a temperature of 70 °C, we apply Equation (7.2) and we get an increase in resistance of 20 per cent as shown in Equation (7.3). The resistance/temperature coefficient at 20 °C for copper is 0.004 per °C:

$$R_t = R_{20}\left[1+\alpha_{20}(t-20)\right] = 2\left[1+0.004(70-20)\right] = 2\left[1+0.2\right] = 2.4 \ (\Omega) \tag{7.3}$$

Equation (7.3) shows that the resistance at 70 °C is 20 per cent higher than it is at 20 °C. Bearing in mind that reactance is unaffected by temperature, this would support the 80 per cent values given for comparison with measured values of earth fault loop impedance, Z_s, as a rule of thumb.

▶ **Table 7.3** Data from Table 41B1 of BS 7671 for maximum earth fault loop impedance (Z_s) for fuses, for 0.4 s disconnection time with U_0 of 230 V (see Regulation 413-02-10)[1], together with 80 per cent values[2] for comparison with measured values taken at an ambient temperature of 20 °C

(a) General purpose (gG) fuses to BS 88-2.1 and BS 88-6								
Rating (A)	6	10	16	20	25	32	40	50
Z_s (Ω)[1]	8.89	5.33	2.82	1.85	1.50	1.09	0.86	0.63
Z_s (Ω) measured[2]	7.11	4.26	2.25	1.48	1.20	0.87	0.68	0.50
(b) Fuses to BS 1361								
Rating (A)	5	15	20	30	45			
Z_s (Ω)[1]	10.9	3.43	1.78	1.20	0.60			
Z_s (Ω) measured[2]	8.72	2.74	1.42	0.96	0.48			
(c) Fuses to BS 3036								
Rating (A)	5	15	20	30	45			
Z_s (Ω)[1]	10.0	2.67	1.85	1.14	0.62			
Z_s (Ω) measured[2]	8.00	2.13	1.48	0.91	0.49			
(d) Fuses to BS 1362								
Rating (A)		13						
Z_s (Ω)[1]		2.53						
Z_s (Ω) measured[2]		2.02						

(e) Type B circuit-breakers to BS EN 60898 and RCBOs to BS EN 61009												
Rating (A)	6	10	16	20	25	32	40	50	63	80	100	125	I_n
Z_s (Ω)[1]	8.00	4.80	3.00	2.40	1.92	1.50	1.20	0.96	0.76	0.60	0.48	0.38	$48/I_n$
Z_s (Ω) measured[2]	6.40	3.84	2.40	1.92	1.53	1.20	0.96	0.76	0.60	0.48	0.38	0.30	
(f) Type C circuit-breakers to BS EN 60898 and RCBOs to BS EN 61009												
Rating (A)	6	10	16	20	25	32	40	50	63	80	100	125	I_n
Z_s (Ω)[1]	4.00	2.40	1.50	1.20	0.96	0.75	0.60	0.48	0.38	0.30	0.24	0.19	$24/I_n$
Z_s (Ω) measured[2]	3.20	1.92	1.20	0.96	0.76	0.60	0.48	0.38	0.30	0.24	0.19	0.15	
(g) Type D circuit-breakers to BS EN 60898 and RCBOs to BS EN 61009												
Rating (A)	6	10	16	20	25	32	40	50	63	80	100	125	I_n
Z_s (Ω)[1]	2.00	1.20	0.75	0.60	0.48	0.38	0.30	0.24	0.19	0.15	0.12	0.10	$12/I_n$
Z_s (Ω) measured[2]	1.60	0.96	0.60	0.48	0.38	0.30	0.24	0.19	0.15	0.12	0.09	0.08	

▶ **Table 7.4** Data from Table 41B2 of BS 7671 for maximum earth fault loop impedance (Z_s) for circuit-breakers with U_0 of 230 V, or instantaneous operation giving compliance with the 0.4 s disconnection time of Regulation 413-02-11 and the 5 s disconnection time of Regulations 413-02-12 and 413-02-14[1], together with 80 per cent values[2] for comparison with measureed values taken at an ambient temperature of 20 °C

(a) General purpose (gG) fuses to BS 88-2.1 and BS 88-6							
Rating (A)	6	10	16	20	25	32	40
Z_s (Ω)[1]	14.1	7.74	4.36	3.04	2.40	1.92	1.41
Z_s (Ω) measured[2]	11.28	6.19	3.48	2.43	1.92	1.53	1.12
Rating (A)	50	63	80	100	125	160	200
Z_s (Ω)[1]	1.09	0.86	0.60	0.44	0.35	0.27	0.20
Z_s (Ω) measured[2]	0.87	0.68	0.48	0.35	0.28	0.21	0.16
(b) Fuses to BS 1361							
Rating (A)	5	15	20	30	45	60	80	100
Z_s (Ω)[1]	17.1	5.22	2.93	1.92	1.00	0.73	0.52	0.38
Z_s (Ω) measured[2]	13.68	4.17	2.34	1.53	0.80	0.58	0.41	0.30
(c) Fuses to BS 3036							
Rating (A)	5	15	20	30	45	60		100
Z_s (Ω)[1]	18.5	5.58	4.00	2.76	1.66	1.17		0.56
Z_s (Ω) measured[2]	14.80	4.46	3.20	2.20	1.32	0.93		0.44
(d) Fuses to BS 1362							
Rating (A)	13						
Z_s (Ω)[1]	4.00						
Z_s (Ω) measured[2]	3.2						

▶ **Table 7.5** Data from Table 41D of BS 7671 for maximum earth fault loop impedance (Z_s) for 5 s disconnection time with U_0 of 230 V (see Regulations 413-02-13 and 413-02-14)[1], together with 80 per cent values[2] for comparison with measured values taken at an ambient temperature of 20 °C

Table 7.6 Data from Table 604B1 of BS 7671 for maximum earth fault loop impedance (Z_s) for fuses, for 0.2 s disconnection time with U_0 of 230 V (see Regulation 604-04-03)[1], together with 80 per cent values[2] for comparison with measured values taken at an ambient temperature of 20 °C

(a) General purpose (gG) fuses to BS 88-2.1 and BS 88-6								
Rating (A)	6	10	16	20	25	32	40	50
Z_s (Ω)[1]	7.74	4.71	2.53	1.60	1.33	0.92	0.71	0.53
Z_s (Ω) measured[2]	6.19	3.76	2.02	1.28	1.06	0.73	0.56	0.42
(b) Fuses to BS 1361								
Rating (A)	5	15	20	30	45			
Z_s (Ω)[1]	9.60	3.00	1.55	1.00	0.51			
Z_s (Ω) measured[2]	7.68	2.40	1.24	0.80	0.40			
(c) Fuses to BS 3036								
Rating (A)	5	15	20	30	45			
Z_s (Ω)[1]	7.50	1.92	1.33	0.80	0.41			
Z_s (Ω) measured[2]	6.00	1.53	1.06	0.64	0.32			
(d) Fuses to BS 1362								
Rating (A)			13					
Z_s (Ω)[1]			2.14					
Z_s (Ω) measured[2]			1.71					

For a circuit with a nominal voltage other than 230 V, it is important to recognise that for the operation of an overcurrent protective device, such as a fuse or circuit-breaker, it is the same magnitude of the earth fault current that is required. Therefore, taking a practical example of a circuit with a nominal voltage of 115 V (U_0) protected by a 16 A, BS 88-6 fuse supplying socket-outlets, the limit on the earth fault loop impedance (Z_s) would be the value given in Table 41B of BS 7671 (2.82 Ω) multiplied by a factor of 115/230 (½), representing the ratio of two different values of U_0. The limit on Z_s would be 1.41 Ω.

Under certain circumstances the disconnection times for final circuits which supply socket-outlets, portable equipment intended for manual movement during use, and hand-held Class I equipment may be increased to a value not exceeding 5 s (see Regulation 413-02-12). This is permitted only where additional requirements on the maximum impedance of the CPC are met. The maximum CPC impedance is given in Table 41C of BS 7671, and for convenience the data given in Table 41C is replicated in Table 7.7 for fuses and Table 7.8 for circuit-breakers.

The earth fault loop can be seen to have a number of discrete constituent parts which all contribute to the earth fault loop impedance, as shown in Figure 7.2, from which the principal parts shown are:

- source impedance
- supply phase conductor
- installation phase conductor
- circuit protective conductor
- earthing conductor
- PEN conductor of the supply.

For the general case, the earth fault loop impedance (Z_s) is as given in Equation (7.4):

$$Z_s = Z_e + (Z_1 + Z_2) \ (\Omega) \tag{7.4}$$

where: Z_e is the external earth fault loop impedance (made up principally of the source impedance, the supply phase conductor and the PEN conductor of the supply),
Z_1 is the impedance of the installation phase conductor to the point of the fault, and
Z_2 is the impedance of the CPC.

By convention, the impedance of the fault is taken to be negligible.

For circuits rated at less than 100 A and operating at a supply frequency not exceeding 50 Hz, the inductive reactance can be ignored. Therefore, for most final circuits, the impedances Z_1 and Z_2 may be replaced by the resistive components of impedance R_1 and R_2, the resistances of the phase and CPCs, as given in Equation (7.5):

$$Z_s = Z_e + (R_1 + R_2) \ (\Omega) \tag{7.5}$$

where: Z_e is the external earth fault loop impedance (made up principally of the source impedance, the supply phase conductor and the PEN of the supply),
R_1 is the resistance of the installation phase conductor to the point of the fault, and R_2 is the resistance of the CPC.

To make certain that automatic disconnection occurs within the specified time, the value of Z_s at the furthest (the electrically most remote) point of a circuit of nominal voltage to earth (U_0) of 230 V should not exceed the limiting value given in Tables 41B1, 41B2 and 41D of BS 7671 (Tables 7.3, 7.4, 7.5 and 7.6 of this Guide) for the selected protective device and for the prescribed limit on disconnection time. The limiting values given in the tables assume that the conductors are at their normal operating temperature. Where the conductors are at a different temperature when the values of the impedances or resistances are established (normally by measurement), the readings should be adjusted accordingly.

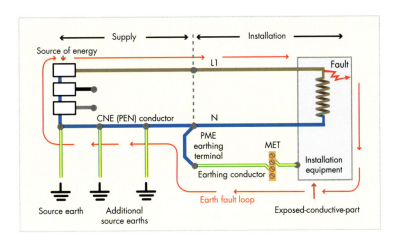

▶ **Figure 7.2** Earth fault loop (TN-C-S system)

7.2.2 Mixed disconnection times

As recognised by Regulation 413-02-13, where circuits with differing disconnection times emanate from a common distribution board, or consumer unit, there is a risk that under earth fault conditions the touch voltage on an exposed-conductive-part of a circuit having a disconnection time of 5 s will be imposed on exposed-conductive-parts of a circuit where 0.4 s disconnection is required. Figure 7.3 illustrates this situation.

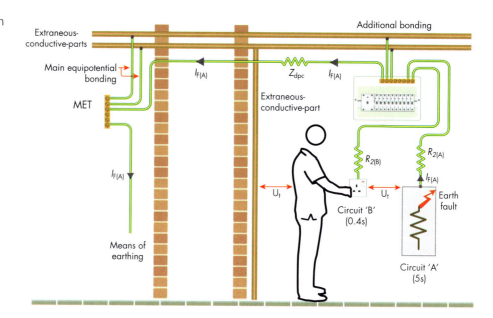

▶ **Figure 7.3** Circuits with differing disconnection times

Take the installation shown in Figure 7.3 which consists of a 230 V distribution board that is remote from the origin of the installation, and two final circuits, circuit A and circuit B. Only CPCs and main equipotential bonding conductors are shown to aid clarity.

The impedances of the circuit protective conductors are identified thus:

▶ Z_{dpc} relates to the CPC of the distribution circuit serving the distribution board
▶ R_{2A} is the CPC of circuit A which emanates from the distribution board
▶ R_{2B} is the CPC of circuit B which emanates from the distribution board.

Circuit A feeds an item of Class I fixed equipment and has a maximum permitted disconnection time of 5 s.

Circuit B feeds a socket-outlet and has a maximum permitted disconnection time of 0.4 s.

I_{FA} is the earth fault current caused by an earth fault in fixed equipment fed by circuit A.

A fault between a phase conductor and an exposed-conductive-part occurring on circuit A will result in a touch voltage (potential difference), between the distribution board earth bar and extraneous-conductive-parts. The voltage arises from the fault current passing through the distribution board's CPC, and will be of a magnitude equal to the product of the impedance and the earth fault current, as shown in Equation (7.6):

$$U_t = I_{FA} \times Z_{dpc} \ (V) \tag{7.6}$$

The touch voltage, U_t, will not only exist between the distribution board's earthing bar and extraneous-conductive-parts, it will also exist between those parts and the socket-outlet on circuit B and any exposed-conductive-parts of Class I equipment to which it is connected. This exported touch voltage would be maintained for the duration of the earth fault which may be up to 5 s.

What the requirements of Regulation 413-02-13 for circuits with mixed disconnection times set out to do is to limit the magnitude of the exported voltage to a value which, when persisting for the disconnection time of circuit A, will not cause danger.

Whilst the requirements of Regulation 413-02-13 concerning mixed disconnection times relate to final circuits emanating from the same distribution board or distribution circuit, the situation in which a number of distribution boards are located together and linked electrically by earthed metalwork (either deliberately or fortuitously) should also be considered in the same light.

Where circuits with mixed disconnection times are connected to the same distribution board or distribution circuit, one of the following conditions is required to be fulfilled, as required by Regulation 413-02-13:

▸ the CPC impedance between the distribution board and the point at which the protective conductor is connected to the main equipotential bonding is limited, or
▸ additional equipotential bonding is provided at the distribution board.

Taking the first of the two options of limiting the impedance of the protective conductor between the distribution board and the point of connection with the main equipotential bonding, shown as Z_{dpc} in Figure 7.3, Regulation 413-02-13(i) requires that the impedance does not exceed the value given in Table 41C of BS 7671 appropriate to the type of protective device used in the final circuit. For convenience, the data given in Table 41C of BS 7671 is replicated here in Table 7.7 (for fuses) and Table 7.8 (for circuit-breakers).

▸ **Table 7.7** Maximum impedance of CPC related to the final circuit protective device of fuses (see Regulation 413-02-12)

		Rating (amperes)						
a	General purpose (gG) fuses to BS 88-2.1 and BS 88-6	6	10	16	20	25	32	40	50
	Limiting impedance (Ω)	2.48	1.48	0.83	0.55	0.43	0.34	0.26	0.19
b	Fuses to BS 1361	5	15	20	30	45			
	Limiting impedance (Ω)	3.25	0.96	0.55	0.36	0.18			
c	Fuses to BS 3036	5	15	20	30	45			
	Limiting impedance (Ω)	3.25	0.96	0.63	0.43	0.24			
d	Fuses to BS 1362	13							
	Limiting impedance (Ω)	0.75							

Notes:
The protective conductor impedances given in this table should not be exceeded when the conductors are at their normal operating temperature. If the conductors are at a different temperature when tested, the reading should be adjusted accordingly.

▸ **Table 7.8** Maximum impedance of CPC related to the final circuit protective device of circuit-breakers and RCBOs (see Regulation 413-02-12)

	Rating (amperes)											
Type B+RCBO	6	10	16	20	25	32	40	50	63	80	100	125	I_n
Limiting impedance (Ω)	1.67	1.00	0.63	0.50	0.40	0.31	0.25	0.20	0.16	0.12	0.10	0.08	$10/I_n$
Type C+RCBO	6	10	16	20	25	32	40	50	63	80	100	125	I_n
Limiting impedance (Ω)	0.83	0.5	0.31	0.25	0.20	0.16	0.13	0.10	0.08	0.06	0.05	0.04	$5/I_n$
Type D + RCBO	6	10	16	20	25	32	40	50	63	80	100	125	I_n
Limiting impedance (Ω)	0.42	0.25	0.16	0.12	0.10	0.08	0.06	0.05	0.04	0.03	0.03	0.02	$2.5/I_n$

Notes:
All circuit-breakers to BS EN 60898 and residual current breakers with overcurrent devices (RCBOs) to BS EN 61009. The protective conductor impedances given in this table should not be exceeded when the conductors are at their normal operating temperature. If the conductors are at a different temperature when tested, the reading should be adjusted accordingly.

The limiting impedance of any protective device that is not listed in Table 41C of BS 7671 (and Tables 7.7 and 7.8 of this Guide) is required to meet Equation (7.7).

$$Z_L \le \frac{50 Z_s}{U_0} \quad (\Omega) \tag{7.7}$$

where: Z_L is the limiting impedance of the CPC
Z_s is the earth fault loop impedance corresponding to a disconnection time of 5 s
U_0 is the nominal voltage of the circuit.

Where the earth fault loop impedance, Z_s, of the final circuit corresponds to a disconnection time of 5 s, a restriction placed on the impedance of the protective conductor between the distribution board and the point of connection with the main equipotential bonding limits the voltage between these points to 50 V. Where Z_s for the final circuit is less than that corresponding to a disconnection time of 5 s, the voltage will exceed 50 V but the disconnection time will be less than 5 s.

Where the option of limiting the impedance of the protective conductor is not practicable or is otherwise undesirable, the designer has a further option of undertaking additional equipotential bonding in accordance with Regulation 413-02-13(ii) which reads:

> *(ii) there shall be equipotential bonding at the distribution board, which involves the same types of extraneous-conductive-parts as the main equipotential bonding according to Regulation 413-02-02 and is sized in accordance with Regulation 547-02-01.*

To meet the above Regulation requires additional equipotential bonding to be carried out at a distribution board remote from the origin of the installation to all extraneous-conductive-parts with bonding conductors selected to have a similar CSA as for main bonding.

7.2.3 Automatic disconnection for socket-outlets etc. in 5 s

Irrespective of the nominal voltage of the circuit, U_0, Regulation 413-02-12 permits the limiting disconnection time for a circuit supplying a socket-outlet or portable equipment intended for manual movement during use, or hand-held Class I equipment to be extended up to 5 s where certain conditions are met, the conditions being:

- Only circuits with the types and ratings of the overcurrent protective devices and associated maximum impedances of the CPCs shown in Table 41C.
- The impedance of the protective conductor should be referred to the point of connection to the main equipotential bonding and should not exceed the values given in Table 41C of BS 7671 appropriate to the particular overcurrent protective device.
- Where additional equipotential bonding is installed in accordance with Regulation 413-02-13(ii) then the impedance of the CPC specified in this paragraph applies to that portion of the CPC between the point of additional bonding and the socket-outlet, or portable equipment.

For a circuit protected by a circuit-breaker, the data relating to the limiting earth fault loop impedance given in Tables 41B2 of BS 7671 for circuit-breakers that comply with BS EN 60898 is for both 5 and 0.4 s disconnection, and the option for extending the disconnection time to 5 s is not therefore of any relevance.

7.2.4 Automatic disconnection using an RCD

An RCD may be used to provide protection against indirect contact in TN systems where the earth fault loop impedance (Z_s) is insufficiently low to operate an overcurrent protective device (fuse or circuit-breaker) within the prescribed disconnection time.

Where an RCD is used to provide automatic disconnection, Regulation 413-02-16 requires Equation (7.8) to be met:

$$Z_s I_{\Delta n} \leq 50 \text{ V} \tag{7.8}$$

where: Z_s is the earth fault loop impedance (Ω)

$I_{\Delta n}$ is the rated residual operating current of the protective device (A)

For circuits extending outside the earthed equipotential zone, Regulation 413-02-17 permits the exposed-conductive-parts to be connected to a separate earth electrode, thereby making the installation part of a TT system with the particular requirements for such systems having to be met.

Where more than one residual current device is used in series the upstream device needs to be time-delayed or S type and the downstream a type for general use, both to BS EN 61008 or BS EN 61009 to achieve discrimination. This is addressed in Clause 7.7.

7.3 TT systems

For installations forming part of a TT system the limits on automatic disconnection are not given in terms of time. However, the limits on disconnection time can be inferred as 5 s by reference to Equation (7.9), which is required to be met by Regulation 413-02-20:

$$R_A I_a \leq 50 \text{ V} \tag{7.9}$$

where: R_A is the sum of the resistances of the earth electrode and the protective conductor(s) connecting it to the exposed-conductive-part, and

I_a is the current causing the automatic operation of the protective device within 5 s. Where the protective device is a residual current device, I_a is the rated residual operating current $I_{\Delta n}$.

Additionally, Regulation 413-02-18 requires all exposed-conductive-parts that are protected against indirect contact by a single RCD be connected to a common earth electrode, via the MET.

Whilst overcurrent protective devices are acceptable as a means of automatic disconnection in installations forming part of a TT system, the earth fault loop impedance is often insufficiently low for their widespread use.

7.4 IT systems

The source of an IT system is either isolated from earth or, where necessary to reduce overvoltage or to damp voltage oscillations, earthed through a high impedance.

Disconnection in the event of a first line fault to earth is not essential, but precautions must be taken to safeguard against the risk of electric shock in the event of two faults existing simultaneously. All exposed-conductive-parts are required to be earthed and the condition expressed in Regulation 413-02-23 and in Equation (7.10) should be met:

$$R_A I_d \leq 50 \text{ V} \tag{7.10}$$

where: R_A is the sum of the resistances of the earth electrode and the protective conductor connecting it to the exposed-conductive-parts, and
I_d is the fault current of the first fault of negligible impedance between a phase conductor and an exposed-conductive-part.

The fault current I_d takes account of protective conductor currents and the total earthing impedance of the electrical installation. As called for in Regulation 413-02-24, it is necessary to provide an insulation monitoring device capable of providing a visual and/or audible warning of the occurrence of a first fault from a live part to an exposed-conductive-part or to earth.

On the occurrence of a second fault, the system should be treated as if it were a TN system or a TT system, depending on how the exposed-conductive-parts are earthed:

▶ where the exposed-conductive-parts are earthed collectively, the conditions for a TN system should apply, but subject to Regulation 413-02-26 and Equation (7.9) or (7.10) being met
▶ where exposed-conductive-parts are earthed in groups or individually, conditions for TT systems apply and (7.9) are required to be met.

For cases in which the neutral is not distributed (three-phase three-wire distribution), Equation (7.11) applies. When the neutral is distributed (three-phase four-wire distribution and single-phase distribution), Equation (7.12) applies:

$$Z_s \leq \frac{\sqrt{3}}{2} \frac{U_0}{I_a} \ (\Omega) \tag{7.11}$$

$$Z_s^1 \leq \frac{1}{2} \frac{U_0}{I_a} \ (\Omega) \tag{7.12}$$

where: Z_s is the impedance of the earth fault loop comprising the phase conductor and the protective conductor of the circuit,
Z_s^1 is the impedance of the earth fault loop comprising the neutral conductor and the protective conductor of the circuit, and
I_a is the current which disconnects the circuit within the time t specified in Table 41E of BS 7671 (replicated here as Table 7.9) where applicable for socket-outlet circuits, or within 5 s for all other circuits where this time is allowed (see Regulation 413-02-13).

▶ **Table 7.9** Data from Table 41E of BS 7671 for maximum disconnection time in IT systems (second fault)

Installation nominal voltage U_0/U (V)	Maximum disconnection time t (s)	
	Neutral not distributed*	Neutral distributed
120/240	0.8	5.0
230/400	0.4	0.8
400/690	0.2	0.4
580/1000	0.1	0.2

Notes:
U_0 is the nominal a.c. rms voltage between phase and neutral.
U is the nominal a.c. rms voltage between phases.
* phase to phase voltages only are available.

Both overcurrent protective devices and residual current protective devices are permitted to be used for automatic disconnection. However, where protection against indirect contact is provided by an RCD, each final circuit is required to be separately protected.

7.5 Automatic disconnection for portable equipment for use outdoors

For all socket-outlets and other circuits that may reasonably be expected to supply portable equipment for use outdoors, supplementary protection should be provided as required by Regulation 471-16-01, which for convenience is reproduced here.

> *A socket-outlet rated 32 A or less which may reasonably be expected to supply portable equipment for use outdoors shall be provided with supplementary protection to reduce the risk associated with direct contact by means of a residual current device having the characteristics specified in Regulation 412-06-02(ii).*
>
> *This Regulation does not apply to a socket-outlet supplied by a circuit incorporating one or more of the protective measures specified in items (i) to (iii) below and complying with the Regulations indicated:*
>
> *(i) Protection by SELV (see Regulation Groups 411-02 and 471-02)*
>
> *(ii) Protection by electrical separation (see Regulations 413-06 and 471-12)*
>
> *(iii) Protection by automatic disconnection and reduced low voltage systems (see Regulation Group 471-15).*

Similar requirements apply to a circuit supplying portable equipment for use outdoors, connected other than through a socket-outlet by means of flexible cable or cord having a current-carrying capacity of 32 A or less, as required by Regulation 471-16-02.

An RCD selected for this supplementary protection against direct contact is required to have a rated residual current, $I_{\Delta n}$, of not more than 30 mA and an operating time not exceeding 40 ms when subjected to a residual current of five times $I_{\Delta n}$ as provided for in product standards BS 4293, BS 7071, BS 7288, BS EN 61008-1 and BS EN 61009-1.

7.6 Automatic disconnection for circuits supplying fixed equipment outdoors

As called for in Regulation 471-08-03, for a circuit which supplies fixed equipment outside the earthed equipotential zone and has exposed-conductive-parts which may be touched by a person directly in contact with the general mass of Earth, the disconnection time is limited to that given in Table 41A of BS 7671. For a circuit supplied at a nominal voltage, U_0, of 230 V, the maximum disconnection time is given as 0.4 s.

7.7 RCDs in series

It is often necessary to employ RCDs in series or cascade as for example in an installation forming part of a TT system. Regulation 531-02-09 of BS 7671 requires that, where it is necessary to prevent danger, the characteristics of the RCDs are such that discrimination is achieved. Even where it is not strictly necessary to provide discrimination to prevent danger or to minimise inconvenience (Regulation 314-01-01), it is considered to be good practice to do so.

For two RCDs in series, discrimination will generally be achieved by selecting a time-delayed characteristic for the upstream RCD. Figure 7.4 shows two RCDs in series feeding an item of Class I current-using equipment which has developed an earth fault.

▶ **Figure 7.4** RCDs in series feeding an item of Class I current-using equipment

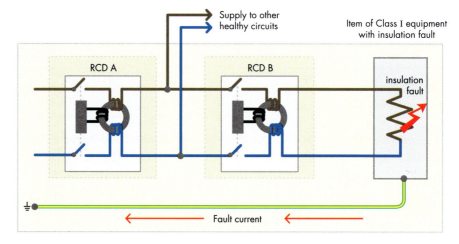

As a consequence of an earth fault, an earth fault current, I_{fault}, flows through the phase conductor and on to earth via the protective conductor. This earth fault current, by flowing in the protective conductor, causes an imbalance in the currents in the live conductors. This imbalance is sensed in both RCD A and RCD B and, where both devices do not have any time-delay mechanism, both devices are likely to operate.

This dual operation is likely even where there is a significant difference in the rated residual currents of the devices, and power could be lost to healthy circuits as well as to the circuit with the earth fault.

It is important to recognise that an RCD does not limit the earth fault current, which is a function of the nominal voltage to earth, U_0, and earth fault loop impedance, Z_s, as given in Equation (7.13):

$$I_{fault} = \frac{U_0}{Z_s} \ (A) \tag{7.13}$$

Unless the earth fault loop impedance, Z_s, is unusually high (and probably unacceptably high) the earth fault current will exceed that required to operate both devices. Take, for example, a situation in which the earth fault loop impedance is 100 Ω (TT system). The resulting earth fault current, I_{fault}, would be 2.3 A as shown by Equation (7.14):

$$I_{fault} = \frac{U_0}{Z_s} = \frac{230}{100} = 2.3 \ A \tag{7.14}$$

An earth fault current of 2.3 A is likely to exceed the rated residual operating currents of both devices and therefore both will operate. To discriminate in terms of magnitude only is not therefore a plausible option in most cases.

Discrimination between RCDs can generally only be achieved by 'forcing' a delay in the operation of the upstream RCD, thus allowing the RCD downstream time to disconnect the fault before the time delay of the upstream RCD has been exhausted.

Where an RCD is provided for supplementary protection against direct contact, a time-delay mechanism is not permitted. Such devices are required to meet the type-test disconnection time laid down in the product standard, as required by Regulation 412-06-02.

Where an RCD is provided for protection against indirect contact, and does have a time-delay mechanism, it is required to automatically disconnect within the specified overall time limit for the particular circuit(s) and the particular location.

7.8 Automatic disconnection for reduced low voltage circuits

An automatic disconnection and reduced low voltage system is often used where for functional reasons the use of PELV (protective extra-low voltage) is impracticable and SELV is not necessary.

BS 7671 defines a reduced low voltage system as:

> *A system in which the nominal phase to phase voltage does not exceed 110 volts and the nominal phase to earth voltage does not exceed 63.5 volts.*

Although the source may be a motor-generator or an engine-driven generator, the most common source is a double-wound isolating transformer complying with BS 3535 (replaced by BS EN 60742).

Figure 7.5 shows the secondary windings of both a three-phase and a single-phase source.

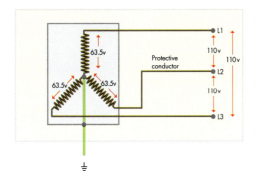
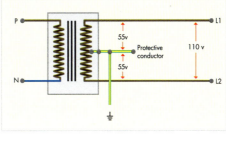

▶ **Figure 7.5** Sources for reduced low voltage systems

As required by Regulation 471-15-06, protection against indirect contact is to be provided by automatic disconnection. It is not reliant on earthed equipotential bonding and for this reason it is commonly a preferred protective measure for hand-held tools and small mobile plant on construction and demolition sites. However, it is important that all exposed-conductive-parts are connected to earth.

A reduced low voltage system, which requires the star-point of a three-phase system or the mid-point of a single-phase system to be earthed, is not to be confused with 110 V systems often used with control circuitry which are not so connected. The latter can only be described as low voltage and the general requirements of BS 7671 apply.

Care is required in the selection of a portable generator for a reduced low voltage system. Many 240/110 V single-phase generators do not have a centre-tap available and could therefore only be earthed on one pole. This would not constitute a reduced low voltage system as prescribed in BS 7671, and it would not provide the same degree of protection against indirect contact. It would be a low voltage system.

Regulation 471-15-06 requires that where automatic disconnection is provided by overcurrent protective devices (such as a fuse or a circuit-breaker) a device is required in each phase conductor. Disconnection for indirect contact is required to occur in not more than 5 s, and this applies to every point in the circuit including socket-outlets. For such devices, the earth fault loop impedance, Z_s, is required to meet that given in Equation (7.15):

$$Z_s \leq \frac{U_0}{I_a} \; (\Omega) \tag{7.15}$$

where: Z_s is the earth fault loop impedance,
I_a is the current causing the automatic operation of the disconnecting protective device within a time not exceeding 5 s, and
U_0 is the nominal voltage to earth.

The limiting value of the earth fault loop impedance, Z_s, may be calculated using Equation (7.15). Alternatively, Z_s can be determined from Table 471A of BS 7671, the data of which is replicated here as Table 7.10.

▶ **Table 7.10** Table 471A of BS 7671 - Maximum earth fault loop impedance (Z_s ohms) for a disconnection time of 5 s and U_0 of 55 V (single-phase) or 63.5 V (three-phase) (see Regulations 471-15-02 and 471-15-06)

	Circuit-breakers to BS EN 60898 and RCBOs to BS EN 61009						General purpose (gG) fuses to BS 88-2.1 and BS 88-6	
	Type B		Type C		Type D			
U_0 (V)	55	63.5	55	63.5	55	63.5	55	63.5
Rating (A)								
6	1.83	2.12	0.92	1.07	0.47	0.53	3.20	3.70
10	1.10	1.27	0.55	0.64	0.28	0.32	1.77	2.05
16	0.69	0.79	0.34	0.40	0.18	0.20	1.00	1.15
20	0.55	0.64	0.28	0.32	0.14	0.16	0.69	0.80
25	0.44	0.51	0.22	0.26	0.11	0.13	0.55	0.63
32	0.34	0.40	0.17	0.20	0.09	0.10	0.44	0.51
40	0.28	0.32	0.14	0.16	0.07	0.08	0.32	0.37
50	0.22	0.25	0.11	0.13	0.06	0.06	0.25	0.29
63	0.17	0.20	0.09	0.10	0.04	0.05	0.20	0.23
80	0.14	0.16	0.07	0.08	0.04	0.04	0.14	0.16
100	0.11	0.13	0.05	0.06	0.03	0.03	0.10	0.12
125	0.09	0.10	0.04	0.05	0.02	0.03	0.08	0.09
I_n	11 I_n	12.7 I_n	5.5 I_n	6.4 I_n	2.8 I_n	3.2 I_n		

Notes:
The circuit loop impedances given in the table should not be exceeded when the conductors are at their normal operating temperature. If the conductors are at a different temperature when tested, the reading should be adjusted accordingly.

Where an RCD is used to provide automatic disconnection, Regulation 471-15-06 requires that the product of the rated residual operating current in amperes, $I_{\Delta n}$, and the earth fault loop impedance, Z_s, in ohms does not exceed 50 V, as expressed in Equation (7.16):

$$I_{\Delta n} \times Z_s \leq 50 \text{ V} \qquad (7.16)$$

The earth fault loop impedance, Z_s, may be assessed by the use of Equation (7.17).

$$Z_s = Z_p \times \left(\frac{V_s}{V_p}\right)^2 + \left(\frac{Z\% \text{ Trans} \times V_s^2}{100 \times VA}\right) + (R_1 + R_2)_s \qquad (7.17)$$

where: Z_p is the loop impedance of the primary circuit including that of the source of supply, Z_e
$Z\%$ Trans is the percentage impedance of the step-down transformer
VA is the rating of the step-down transformer
V_s is the secondary voltage
V_p is the primary voltage
$(R_1 + R_2)_s$ are the secondary circuit phase and protective conductor resistances

Where data on the step-down transformer is not available, Equation (7.17) may be simplified to the following:

$$Z_s = 1.25 \left(Z_p \times \left(\frac{V_s}{V_p}\right)^2 + (R_1 + R_2)_s \right) \qquad (7.18)$$

$$Z_s = 1.25 \left(\left[(Z_e + (2R_1)_p) \times \left(\frac{V_s}{V_p}\right)^2 \right] + (R_1 + R_2)_s \right) \qquad (7.19)$$

where: 1.25 is a factor to compensate for an underestimate
Z_e is the external phase-neutral/earth loop impedance
$(2R_1)_p$ is the primary circuit phase plus neutral conductor impedance

7.9 Automatic disconnection and alternative supplies

An important aspect of providing protection against indirect contact, which can be readily overlooked by the designer, is the need to ensure satisfactory operation of the relevant protective device(s) when the installation, or part thereof, is energised from a second source. This may be an alternative low voltage feed, a standby generator supply or an uninterruptible power supply (UPS). To be certain that the requirements of BS 7671 for protection against electric shock (and short-circuits) will still be satisfied, the designer must obtain full information for the alternative or special supply and make the necessary checks on his/her design, which may have been based only upon the characteristics of the normal supply source.

A generator control panel or UPS equipment may include self-protection, a feature of which is the rapid collapse of the output voltage to the load. This will inhibit the operation of any fault protective device situated beyond the equipment terminals and the feature cannot be assumed to provide a fail-safe operational arrangement for the user. Safety of the system as a whole should be ensured by, if necessary, involving the equipment supplier.

7.10 Separated extra-low voltage systems (SELV)

A SELV is an extra-low voltage system which is electrically separated from earth and from other systems in such a way that a single fault cannot give rise to the risk of electric shock. It should be appreciated that if a single fault were to occur it is intended that this should be confined to the SELV system and not involve any conductive parts of another system. Thus, the construction of a SELV system necessitates the use of high-integrity equipment and materials. Figure 7.6 shows the configuration of a SELV source from which it is noted that the secondary winding has no connection with earth.

▶ **Figure 7.6** A SELV source

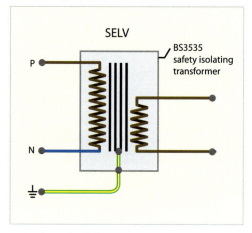

Because a SELV system is devoid of the connection of any of its parts to earth and is itself a protective measure against indirect contact, a means of automatic disconnection is unnecessary from a viewpoint of electric shock although it may be necessary for overcurrent protection.

7.11 Protective extra-low voltage systems (PELV)

The degree of safety afforded by a SELV system depends crucially upon it being isolated both from Earth and from any other system. If this cannot be achieved and maintained throughout its life, the extra-low voltage system cannot depend solely upon the precautions for SELV.

A PELV is an extra-low voltage system which is not electrically separated from earth, but otherwise satisfies all the other requirements for a SELV. Figure 7.7 shows the configuration of a PELV source from which it is noted that the secondary winding has a connection with earth.

▶ **Figure 7.7** A PELV source

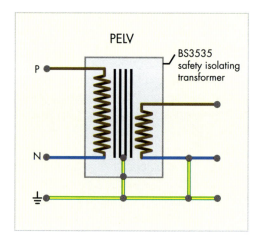

PELV systems rely for protection against indirect contact faults introduced from the primary circuit, on the primary circuit protection. It should be confirmed that this protection against indirect contact is appropriate for the location.

Faults elsewhere on the installation will introduce fault voltages into the PELV system via the protective conductor.

7.12 Functional extra-low voltage systems (FELV)

FELV (functional extra-low voltage) is used where extra-low voltage is required for functional purposes, such as machine control systems. Figure 7.8 shows the configuration of a FELV source from which it is noted that the secondary winding has a connection with earth.

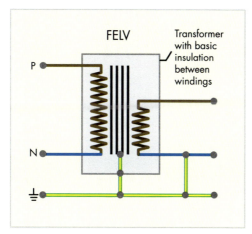

▶ **Figure 7.8** A FELV source

Protection against indirect contact is provided by connecting the exposed-conductive-parts of the FELV circuit to the protective conductor of the primary circuit, provided that the primary circuit is protected by earthed equipotential bonding and automatic disconnection of the supply.

Protection against indirect contact by automatic disconnection of supply in the event of a fault to earth on the secondary side of the system is not a requirement.

Supplementary equipotential bonding 8

8.1 Supplementary equipotential bonding

Supplementary equipotential bonding is a part of the EEBADS protective measure that is used for the protection against indirect contact required by BS 7671 in certain situations and locations. Supplementary equipotential bonding involves connecting together the exposed-conductive-parts and extraneous-conductive-parts such as those shown in Figure 8.1, to minimise the touch voltages between them under earth fault conditions of a circuit. The figure shows conceptually both main equipotential bonding and supplementary equipotential bonding. It is important to note that main equipotential bonding is always required in addition to supplementary equipotential bonding where the protective measure is EEBADS.

Where necessary, supplementary equipotential bonding is applied to an installation or location in order to establish an equipotential zone within that particular part of the installation or location. It has the effect of re-establishing the equipotential reference at that location for all the exposed-conductive-parts and all extraneous-conductive-parts which are required to be bonded together locally. This further reduces any potential differences that may arise between any of these parts during an earth fault.

Supplementary equipotential bonding is required by BS 7671 to be provided in the following circumstances:

▶ installations and locations of increased shock risk, some of which are addressed in Part 6 of BS 7671
▶ where, in the event of an earth fault, the conditions for automatic disconnection cannot be fulfilled by an overcurrent protective device (alternatively, automatic disconnection must be provided by an RCD), as envisaged by Regulation 413-02-04.

Where supplementary equipotential bonding is required, compliance with Regulations 413-02-27, 547-03-01 to 547-03-05, and Regulation 413-02-28, as appropriate is necessary.

Figure 8.1 Application of supplementary equipotential bonding in a location

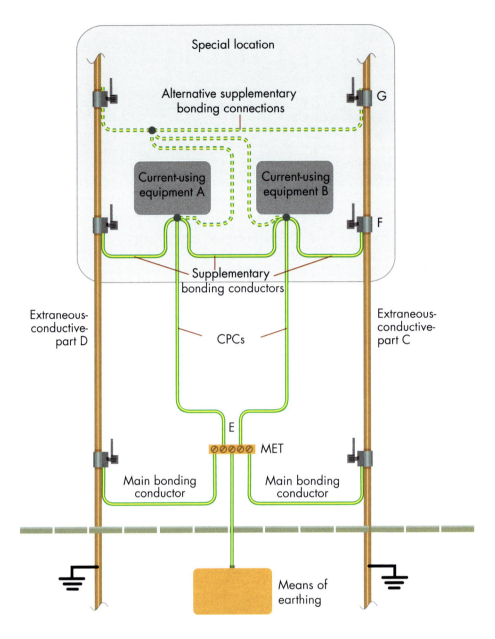

Figure 8.1 shows two items of current-using equipment 'A' and 'B' and the CPCs of the circuits feeding them, together with two separate extraneous-conductive-parts 'C' and 'D'. In accordance with requirements for main equipotential bonding in BS 7671, the CPCs and the extraneous-conductive-parts are connected to the MET of the installation.

Supplementary equipotential bonding is carried out by installing bonding conductors between the exposed-conductive-parts and the extraneous-conductive-parts, by making the connections shown between items of current-using equipment A and B and between B and F. The conductors connecting A with B are required to comply with Regulation 547-03-01 whereas the conductor between B and extraneous-conductive-part C (connection made at point F) is required to comply with Regulation 547-03-02.

The extraneous-conductive-part D is connected either to current-using equipment A or B or to extraneous-conductive-part C, as indicated by the broken lines. Where it is connected to either item of current-using equipment the bonding conductor is required to comply with Regulation 547-03-02. Alternatively, where the connection is made instead to the extraneous-conductive-part C, compliance with Regulation 547-03-03 is required. It should be noted that, as permitted by Regulation 547-03-04, the portion of extraneous-conductive-part C between the points F and G can be considered to be part of the supplementary equipotential bonding.

It is not a requirement of BS 7671 to connect the supplementary bonding conductor back to the MET of the installation, although as shown in Figure 8.1, the locally bonded parts will be connected to this terminal by virtue of one or more CPCs and/or extraneous-conductive-parts.

As previously mentioned, supplementary equipotential bonding may be required in locations where there is considered to be an increased risk of electric shock. Additionally, it may also be required where the conditions for automatic disconnection of Regulation Group 413-02 cannot be fulfilled by an overcurrent protective device. The locations and situations where supplementary equipotential bonding is required by BS 7671 are addressed in Clauses 8.6 to 8.14.

8.2 Supplementary bonding conductor types

Although a single-core, non-flexible, copper cable having a green-and-yellow covering is often used as a supplementary equipotential bonding conductor, there are other types that are also suitable for such use as permitted by Regulations 543-02-02 and 547-03-04, including:

- a protective conductor forming part of a sheathed cable
- metal parts of wiring systems
- extraneous-conductive-parts (e.g. metallic hot and cold water pipes and heating pipework). Note that neither a gas nor an oil pipe may be so used.

Where supplementary equipotential bonding is required to a Class I fixed appliance connected by a short length of flexible cord from an adjacent connection unit or flex outlet accessory, Regulation 547-03-05 permits this to be provided by the CPC within the flexible cord or cable. However, the CPC is required to be connected to the supplementary equipotential bonding within the connection unit or other accessory. Figure 8.2 indicates the application of this regulation.

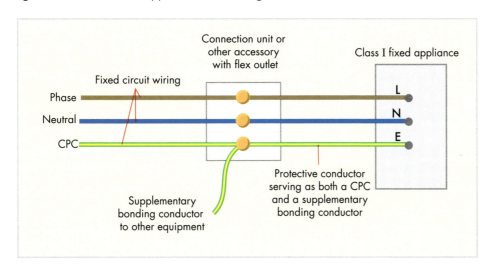

▶ **Figure 8.2**
Supplementary equipotential bonding via a short flexible cord or cable

8.3 CSAs of supplementary bonding conductors

The requirements for the minimum CSA of supplementary equipotential bonding conductors are given in Regulations 547-03-01 to 547-03-03, which are summarised in Table 8.1.

▶ **Table 8.1** Minimum cross-sectional area of supplementary equipotential bonding conductors

Connecting conductive parts	Conductor type	CSA of conductors
Two exposed-conductive-parts (Regulation 547-03-01)	Where sheathed or otherwise provided with mechanical protection	For conductors of the same metal, the bonding conductor to have a conductance not less than that of the smaller CPC connected to the exposed-conductive-parts. For other metals see below
	Where not provided with mechanical protection (e.g. by a sheath or enclosure in conduit etc.)	Not less than 4 mm²
Exposed-conductive-part and an extraneous-conductive-part (Regulation 547-03-02)	Where sheathed or otherwise provided with mechanical protection	For conductor of the same metal, the bonding conductor to have a conductance not less than half that of the protective conductor connected to the exposed-conductive-part. For other metals see below
	Where not provided with mechanical protection (e.g. by a sheath or enclosure in conduit etc.)	Not less than 4 mm²
Two extraneous-conductive-parts (Regulation 547-03-03)	Where neither of the extraneous-conductive-parts is connected to an exposed-conductive-part	Not less than: 2.5 mm² if sheathed or otherwise provided with mechanical protection, or 4 mm² if not provided with mechanical protection

Notes:
1 The insulation of a single-core non-sheathed cable (e.g. to BS 6004 or BS 7211) is not a sheath.
2 Where one of the extraneous-conductive-parts is connected to an exposed-conductive-part, Regulation 547-03-02 is to be applied to the supplementary bonding conductor connecting the two extraneous-conductive-parts (see details earlier in this table).

When the supplementary equipotential bonding conductor referred to in Table 8.1 is made from a different metal to that of the CPC, the minimum CSA is determined by the application of Equation (8.1) (where Regulation 547-03-01 applies – connecting together two exposed-conductive-parts) or Equation (8.2) (where Regulation 547-03-02 applies – connecting together one exposed-conductive-part and one extraneous-conductive-part):

$$S_b \geq S_p \frac{\rho_b}{\rho_p} \ (\text{mm}^2) \tag{8.1}$$

$$S_b \geq \frac{S_p}{2} \frac{\rho_b}{\rho_p} \ (\text{mm}^2) \tag{8.2}$$

where: S_b is the minimum CSA required for the bonding conductor
S_p is the CSA of the CPC
ρ_b is the resistivity (Ωm) for the conductor material of the supplementary bonding conductor
ρ_p is the resistivity (Ωm) for the conductor material of the CPC

8.4 Limitations on resistance of supplementary bonding conductors

Regulation 413-02-28 requires that the resistance, R, of the supplementary equipotential bonding conductor connecting exposed-conductive-parts and extraneous-conductive-parts is limited to the value expressed in Equation (8.3). For some special locations, such as agricultural premises, the resistance is further limited to that given in Equation (8.4):

$$R \leq \frac{50}{I_a} \ (\Omega) \tag{8.3}$$

$$R \leq \frac{25}{I_a} \ (\Omega) \tag{8.4}$$

where: for the case of an overcurrent protective device, I_a is the minimum current which disconnects the circuit within 5 s, whereas for the case of an RCD, I_a is the rated residual operating current $I_{\Delta n}$.

To determine the limiting resistance of supplementary equipotential bonding conductors for a particular location, all the circuits in that location would have to be considered using the value of the highest I_a or $I_{\Delta n}$ in the appropriate formula. For example, given a circuit protected by a 50 A, BS 88-6 fuse with a minimum current (I_a) required to disconnect in 5 s of 220 A and inputting this data into Equation (8.4) (for a wet location), we get Equation (8.5). This gives a limit on the resistance of the supplementary equipotential bonding conductor of 0.11 Ω:

$$R \leq \frac{25}{I_a} = \frac{25}{220} = 0.11 \ \Omega \tag{8.5}$$

Using the conductor resistance of 4 mm^2 copper of 4.61 mΩ/m and with a limit on the resistance of 0.11 Ω, the supplementary equipotential bonding conductor could be as long as 24 m. Similarly, for a 2.5 mm^2 copper conductor at 7.41 mΩ/m and applying the same equation, the limit on resistance would be 0.55 Ω which would limit the conductor length to 15 m.

For most if not all circumstances, this limitation on the resistance R will not result in a need to increase the CSA of the supplementary equipotential bonding conductors above 4 mm^2 copper. However, this constraint will need to be checked for the most onerous circuit for each special location.

8.5 Supports for supplementary bonding conductors

For the case where a supplementary equipotential bonding conductor forms part of a composite cable, such as a separate core, the method of supporting the bonding conductor will be dictated by the type of cable and the manufacturer's installation instructions.

Single-core supplementary equipotential bonding conductors, and similar conductors, are required to be adequately supported, avoiding non-electrical services such as pipework as a means of support, so that they can resist external influences such as mechanical damage etc (see Section 522 of BS 7671), and any anticipated factors likely to result in deterioration (see Regulation 543-03-01).

Table 8.2 provides some guidance in providing adequate support for supplementary equipotential bonding conductors. Figure 8.3 illustrates an example of a supplementary equipotential bonding conductor properly supported.

▶ **Table 8.2** Recommended spacing of supports of single core rigid copper cables for supplementary equipotential bonding conductors

Overall cable diameter, Φ (mm²)	Horizontal spacing (mm)	Vertical spacing (mm)	Comment
Φ ≤ 3	100	150	This data is provided as a guide and may be overridden by the constraints of good workmanship and by visual considerations
3 < Φ ≤ 5	150	200	
5 < Φ ≤ 10	200	250	
10 < Φ ≤ 15	300	350	

▶ **Figure 8.3** An example of supports for a supplementary equipotential bonding conductor

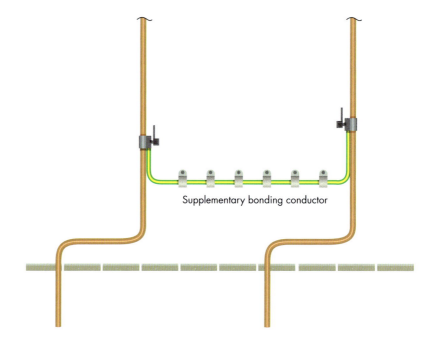

Supplementary bonding conductor

8.6 Bath and shower rooms

Rooms containing a bath and/or shower are locations where there is an increased risk of electric shock due to:

▶ a reduction in body resistance occasioned either by bodily immersion or by wet skin, and
▶ likely contact of substantial areas of the body with the earth potential.

Section 601 in Part 6 of BS 7671: *Special installations or locations* specifically addresses the particular requirements for such locations. It is, however, important to note that these requirements modify or supplement the general requirements contained elsewhere in BS 7671. In other words, compliance with all the requirements of BS 7671 is essential, not just Section 601.

The general requirements for supplementary equipotential bonding are given in Regulations 413-02-27 and 413-02-28. For this location, these Regulations are supplemented by Regulations 601-04-01 and 601-04-02.

The zonal concept applies to this location, and the requirements for supplementary equipotential bonding apply to zones 1, 2 and 3. Regulation 601-04-01 requires the protective conductor terminals of each circuit supplying Class I and Class II electrical equipment in zones 1, 2 or 3, and all extraneous-conductive-parts in these zones to be connected together by local supplementary equipotential bonding conductors complying with Regulation Group 547-03. In so doing, the occurrence of voltages between any such parts of a magnitude sufficient to cause danger of electric shock is prevented. BS 7671 identifies a number of parts which may be considered to be extraneous-conductive-parts:

- metallic pipes supplying services, and metallic waste pipes (e.g. water, gas)
- metallic central heating pipes and air conditioning systems
- accessible metallic structural parts of the building (metallic door architraves, window frames and similar parts are not considered to be extraneous-conductive-parts unless they are connected to metallic structural parts of the building)
- metallic baths and metallic shower basins.

The above listing of extraneous-conductive-parts is not exhaustive and any other parts that fall within the definition of an extraneous-conductive-part given in Part 2 of BS 7671 would require supplementary bonding.

It is also a requirement of Regulation 601-04-01 that the necessary supplementary bonding is required to be carried out either within the location containing the bath or shower or in close proximity. In this context, close proximity is taken to mean in an adjoining roof void or an airing cupboard opening into, or adjoining, the room containing the bath or shower.

Unlike the requirements of previous editions of BS 7671, the requirements for supplementary equipotential bonding in this location apply irrespective of whether or not exposed-conductive-parts and extraneous-conductive-parts are simultaneously accessible. In other words, supplementary equipotential bonding is always required in zones 1, 2 and 3 where exposed-conductive-parts and extraneous-conductive-parts are present.

Regulation 411-02-07 requires that supplementary equipotential bonding is *not* carried out to exposed-conductive-parts of SELV circuits.

Figures 8.4 and 8.5 show sectional views of a typical bathroom, the first with metal pipework and the second with non-metallic pipework.

▶ **Figure 8.4**
Supplementary equipotential bonding in a bathroom – metal pipe installation

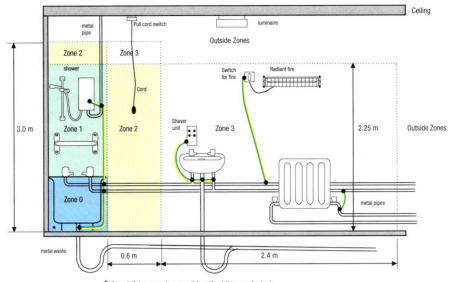

Note 1: The protective conductors of all power and lighting points within the zones must be supplementary bonded to all extraneous-conductive-parts in the zones, including metal waste, water and central heating pipes, and metal baths and metal shower basins.

Note 2: Circuit protective conductors may be used as supplementary bonding conductors.

▶ **Figure 8.5**
Supplementary equipotential bonding in a bathroom – non-metallic pipe installation

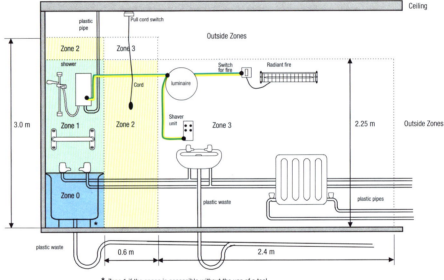

As mentioned earlier, supplementary bonding of the terminals of CPCs of circuits supplying Class II equipment is required as shown in Figures 8.4 and 8.5. This permits Class II equipment to be replaced by Class I equipment during the lifetime of the installation, without the need for further supplementary bonding which might be difficult and costly to carry out subsequently.

For this special location, Regulation 413-02-28 requires that the resistance, R, of the supplementary equipotential bonding connecting exposed-conductive-parts and extraneous-conductive-parts is limited to the value expressed in Equation (8.6).

$$R \leq \frac{50}{I_a} \; (\Omega) \qquad (8.6)$$

where: for the case of an overcurrent protective device, I_a is the minimum current which disconnects the circuit within 5 s, whereas for the case of an RCD, I_a is the rated residual current $I_{\Delta n}$.

Guidance on the CSAs of supplementary equipotential bonding conductors is given in Clause 8.3.

Where electric heating is embedded in the floor of a bathroom or shower room, Regulation 601-09-04 requires that the heating units are covered by an earthed metallic grid or the unit to have an earthed metallic sheath connected in either case to the local supplementary equipotential bonding. Even if the heating units with a metallic sheath are covered by an earthed grid, the heating units are required to be earthed in the normal way.

8.7 Shower cabinet located in a bedroom

Where a cabinet containing a shower is installed in a room other than a bathroom or shower room (such as a bedroom), the supplementary bonding requirements of Regulation 601-04-01 apply only to zones 1 and 2 (Regulation 601-04-02).

Although low voltage socket-outlets (e.g. 230 V) are not permitted in a bath or shower room, they are permitted in a room other than a bath or shower room where a cabinet containing a shower and/or bath is installed. As allowed by Regulation 601-08-02, low voltage socket-outlets may be installed outside zones 0, 1, 2 and 3, provided they are protected by a residual current protective device having a rated residual operating current, $I_{\Delta n}$, not exceeding 30 mA, in accordance with Regulation Group 412-06-02. In cases of retrofit, this requirement can be met by the provision of a socket-outlet with an integral residual current device, as shown in Figure 8.6.

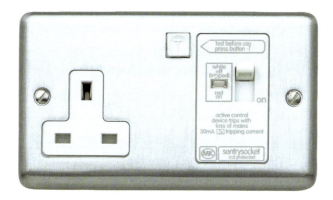

▶ **Figure 8.6** Socket-outlet with RCD ($I_{\Delta n}$ = 30 mA)

8.8 Swimming pools

Swimming pools are locations where there is perceived to be an increased risk of electric shock due to:

▶ a reduction in body resistance occasioned either by bodily immersion or by wet skin, and
▶ likely contact of substantial areas of the body with the earth potential.

Section 602 in Part 6 of BS 7671: *Swimming pools* specifically addresses the particular requirements for such locations. It is, however, important to note that these requirements modify or supplement the general requirements contained elsewhere in BS 7671. In other words, compliance with the entire requirements of BS 7671 is essential, not just Section 602.

As in bath or shower rooms, the zonal concept is used for swimming pools, as shown in Figures 8.7 and 8.8 (which replicate Figures 602A and 602B of BS 7671). In the case of swimming pools, the zones are designated A, B and C (the dimensions for which are given in the figures).

▶ **Figure 8.7** Zone dimensions for swimming pools and paddling pools

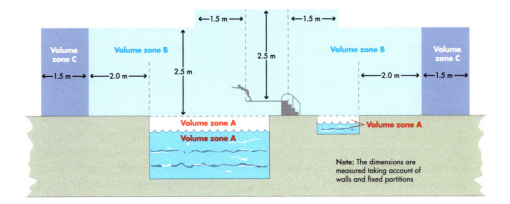

▶ **Figure 8.8** Zone dimensions for a basin above ground

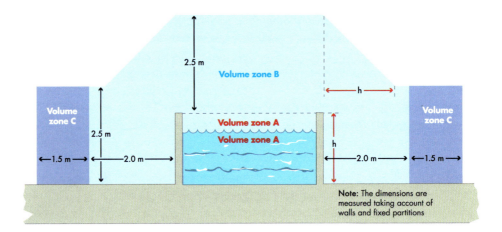

Regulation 602-03-02 requires supplementary equipotential bonding to connect all extraneous-conductive-parts in zones A, B and C with the protective conductors of all exposed-conductive-parts in these zones. However, supplementary equipotential bonding is *not* to be carried out to exposed-conductive-parts of SELV circuits.

Typically, parts which may fall within the definition of an extraneous-conductive-part would include:

▶ exposed structural steelwork
▶ metal pipes etc.
▶ metal handrails.

For this special location, Regulation 413-02-28 requires that the resistance, R, of the supplementary equipotential bonding connecting exposed-conductive-parts and extraneous-conductive-parts is limited to the value expressed in Equation (8.7).

$$R \leq \frac{50}{I_a} \quad (\Omega) \tag{8.7}$$

where: for the case of an overcurrent protective device, I_a is the minimum current which disconnects the circuit within 5 s, whereas for the case of an RCD, I_a is the rated residual current $I_{\Delta n}$.

Guidance on the CSAs of supplementary equipotential bonding conductors is given in Clause 8.3.

Whilst not a requirement of BS 7671, where a metallic grid is installed in a solid floor in zones A, B or C, supplementary equipotential bonding should be provided to connect it with other extraneous-conductive-parts and the protective conductors of all exposed-conductive-parts in the zones.

Where electric heating is embedded in the floor, Regulation 602-08-04 requires that the heating units are covered by an earthed metallic grid or the unit to have an earthed metallic sheath connected in either case to the local supplementary equipotential bonding. Even if the heating units with a metallic sheath are covered by an earthed grid, the heating units are required to be earthed in the normal way.

If the designer considers that a swimming pool floor would fall within the definition of an extraneous-conductive-part, he/she may consider that, to remedy a problem that may at some future time appear, a metallic grid should be installed. The metallic grid is required to have suitable supplementary equipotential bonding. This of course is relatively easy at the site construction stage but highly problematic and costly to retrofit.

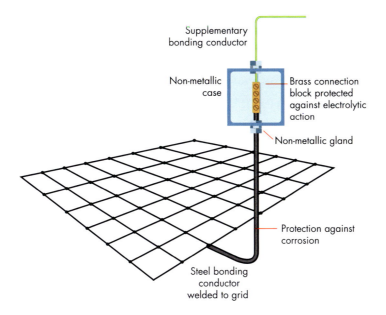

▶ **Figure 8.9** A means of supplementary equipotential bonding a metallic grid

Figure 8.9 illustrates an example of providing supplementary equipotential bonding to a metallic grid. Conductors and connections, which should be accessible for inspection testing and maintenance, should be selected to withstand all the adverse effects of the external influences likely to be present, such as corrosion and mechanical damage.

Where connections cannot be made in accessible positions, such as those made to a metal grid within a concrete slab, they are required, by Regulation 526-04-01, to be made by:

▶ welding
▶ soldering
▶ brazing.

At least two connections should be made to the metallic grid, preferably at two or more diagonally extreme points. Additionally, for the metallic grid to be entirely effective, all constituent parts of the grid should be reliably and durably connected together.

In addition to electric shock as defined, there is also the perception of electric shock to consider in such locations. Electric shock is defined as:

> *A dangerous physiological effect resulting from the passing of an electric current through a human body or livestock.*

The magnitude of the current through the human body which is discernible varies from person to person but it is generally accepted that perception of an electric current can occur from 0.5 mA. With this in mind, the installation designer may decide that the swimming pool installation should form part of a TT system rather than make use of the electricity distributor's earthing facility. In other words, the installation would be earthed to an installation earth electrode rather than to the supply earthing terminal. This may overcome a possible problem with discernible voltages associated in the supply neutral transmitted to the extraneous-conductive-parts of the installation.

However, this solution is not without difficulties. The exposed-conductive-parts and the extraneous-conductive-parts earthed to the installation earth electrode of the special location are required to be separated from parts of the site which are earthed to the distributor's earthing facility. Simultaneous inaccessibility should be maintained at all times.

8.9 Agricultural and horticultural premises

The requirements for agricultural and horticultural premises are given in Section 605 of BS 7671 and apply to all parts of fixed installations of agricultural and horticultural premises outdoors and indoors. They also apply to locations where livestock is kept and to storage areas and the like, such as:

- stables
- chicken-houses
- piggeries
- feed-processing locations
- lofts
- storage areas for hay, straw and fertilizers
- large ventilation ducts.

Dwellings within the curtilage of agricultural and horticultural premises, intended solely for human habitation, are excluded from the scope of this section, although they might have within them requirements for supplementary bonding.

It is important to note that these requirements of Section 605 modify or supplement the general requirements contained elsewhere in BS 7671. In other words, compliance with the entire requirements of BS 7671 is essential, not just Section 605.

Agricultural and horticultural premises are locations where there is perceived to be an increased risk of electric shock to humans and livestock due to:

- a reduction in body resistance occasioned either by partial bodily immersion or by wet skin or hides, and
- likely contact of substantial areas of the body with the earth potential.

The general requirements for supplementary equipotential bonding are given in Regulations 413-02-27 and 413-02-28. These regulations are for this location modified by Regulations 605-08-01 and 605-08-03.

In areas intended for livestock, supplementary equipotential bonding is required to connect together all exposed-conductive-parts and all extraneous-conductive-parts which can be touched by livestock. Typically, this would include milking parlours and associated areas where cows congregate before and after milking. Areas where livestock retire for the night would also be subject to the requirements for supplementary bonding as shown in Figure 8.10.

▶ **Figure 8.10** Cattle at rest [photograph courtesy of Wilson Agriculture]

Livestock containment areas pose a number of challenges for the electrical installation designer. Livestock can be panicked by small sensations of current created by differences in potential between exposed-conductive-parts and extraneous-conductive-parts. Keeping exposed-conductive-parts away from livestock is an obvious objective but this is not always readily achievable. With the proliferation of metal barriers and the like needed to marshal stock within an area, the requirements of supplementary bonding can be demanding. As can be seen from Figure 8.11, there can be a very large number of metal components each of which may require supplementary bonding. Although each metal part would require supplementary equipotential bonding, it would be impractical and unnecessary to take a conductor to each component part where effective and reliable connection is already made by the construction of the metalwork.

Figure 8.11 Metalwork in a typical cowhouse [photograph courtesy of Wilson Agriculture]

Where, for example, the stock areas have concrete slabs (which by their very nature are earthy), it is desirable to lay in a metallic grid within the floor slab to which supplementary equipotential bonding conductors can be connected. Figure 8.12 shows the recommended connection arrangement. This of course is relatively easy at the site construction stage but highly problematic and costly to retrofit.

Figure 8.12 A means of supplementary equipotential bonding to a metallic grid

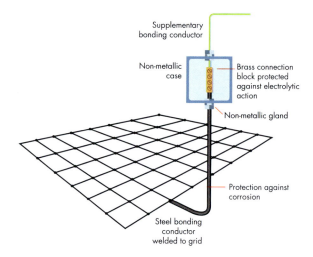

Figure 8.12 illustrates an example of effecting supplementary equipotential bonding to a metallic grid. Conductors and connections, which should be accessible for inspection, testing and maintenance, are required to be selected to withstand all the adverse effects of the external influences likely to be present, such as corrosion and mechanical damage.

Where connections cannot be made in accessible positions, such as those made to a metal grid within a concrete slab, they are required, by Regulation 526-04-01, to be made by:

- welding
- soldering
- brazing.

At least two connections should be made to the metallic grid, preferably at two or more diagonally extreme points. Additionally, for the metallic grid to be entirely effective, all constituent parts of the grid should be reliably and durably connected together.

Regulation 411-02-07 requires that supplementary equipotential bonding is *not* carried out to exposed-conductive-parts of SELV circuits.

For this special location, Regulation 605-08-01 requires that the resistance, *R*, of the supplementary equipotential bonding connecting exposed-conductive-parts and extraneous-conductive-parts is limited to the value expressed in Equation (8.8):

$$R \le \frac{25}{I_a} \; (\Omega) \qquad (8.8)$$

where: for the case of an overcurrent protective device, I_a is the minimum current which disconnects the circuit within 5 s, whereas for the case of an RCD, I_a is the rated residual current $I_{\Delta n}$.

Guidance on the CSAs of supplementary equipotential bonding conductors is given in Clause 8.3.

8.10 Restrictive conductive locations

Section 606 of BS 7671 addresses the particular requirements for this location. However, it is important to note that these requirements of Section 606 modify or supplement the general requirements contained elsewhere in BS 7671. In other words, compliance with the entire requirements of BS 7671 is essential, not just Section 606.

A restrictive conductive location is defined as:

> *A location comprised mainly of metallic or conductive surrounding parts, within which it is likely that a person will come into contact through a substantial portion of their body with the conductive surrounding parts and where the possibility of preventing this contact is limited.*

For example, such a definition may apply to:

- a large boiler interior
- a corn silo
- a large diameter metal pipe or large ventilation duct
- a metal storage tank.

Figure 8.13 illustrates persons gaining access to two examples of restrictive conductive locations.

▶ **Figure 8.13** Access to two examples of restrictive conductive locations

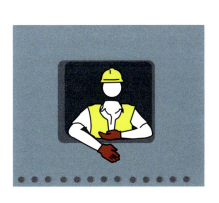

Supplementary equipotential bonding

For protection against indirect contact, four protective measures are permitted by Regulation 606-04-01:

- SELV
- electrical separation
- use of Class II equipment
- automatic disconnection.

It is only the protective measure of automatic disconnection that requires supplementary equipotential bonding to be undertaken in this special location. An additional supplementary bonding conductor is to be provided, meeting the requirements of Regulations 413-02-27 and 413-02-28, connecting the exposed-conductive-parts of the fixed electrical equipment located within the restrictive conductive location and the conductive parts of the location, as shown in Figure 8.14.

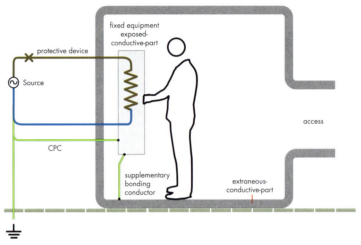

Figure 8.14 Restrictive conductive location showing the required supplementary bonding conductor

Where equipment, such as electronic devices with a protective conductor current, requires a facility for functional earthing, Regulation 606-04-03 requires that supplementary bonding conductors are required to connect together:

- all exposed-conductive-parts of equipment
- all extraneous-conductive-parts inside the restrictive conductive location
- the functional earths of any equipment having protective conductor current.

Figure 8.15 shows the connections of supplementary bonding for functional earthing in a restrictive conductive location.

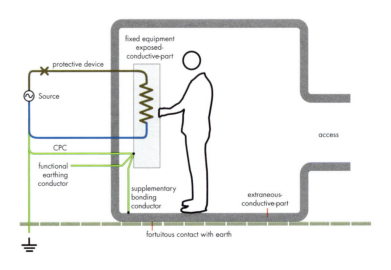

Figure 8.15 Showing the connections of supplementary bonding for functional earthing

8.11 Static inverters

The installation of a static inverter where automatic disconnection cannot be achieved for parts on the load side of the static inverter is required to be provided with supplementary bonding as required by Regulations 413-02-27 and 413-02-28.

As required by Regulation 551-04-04, supplementary equipotential bonding conductors are required to connect together:

- extraneous-conductive-parts, and
- exposed-conductive-parts.

The resistance of the supplementary bonding conductor is limited to the value given in Equation (8.9):

$$R = \frac{50}{I} \quad (\Omega) \tag{8.9}$$

where: R is the resistance of the supplementary bonding conductor, and
I is the maximum fault current which can be supplied by the static inverter alone for a period of up to 5 s.

Where a static inverter is intended to operate in parallel with a distributor's network the requirements of Regulation Group 551-07 also apply.

8.12 Other locations of increased risk

As envisaged in Regulation 471-08-01, there may be locations of increased risk of electric shock other than those specifically addressed in Part 6 of BS 7671.

Where, because of the perception of additional risks being likely the installation designer considers an installation or location warrants further protective measures, the regulation gives three possible options:

- automatic disconnection of supply shall be by means of a residual current device having a rated residual operating current ($I_{\Delta n}$) not exceeding 30 mA
- supplementary equipotential bonding
- reduction of maximum fault clearance time.

The option of supplementary bonding would be required to comply with Regulations 547-03-01 to 547-03-05.

8.13 Where automatic disconnection is not achievable

Regulation 413-02-04 recognises that automatic disconnection within the prescribed time limit with an overcurrent protective device is not always achievable. This might, for example, apply to a circuit with a high current rating where the earth fault loop impedance, Z_s, is too high to cause automatic disconnection within the limit of, say, 5 s.

Taking a practical example of a distribution circuit protected by a 200 A, BS 88-6 fuse for which the earth fault loop impedance at the most remote end of the circuit is determined to be 0.30 Ω, such a value of Z_s is not sufficiently low to cause automatic disconnection within the limit of 5 s. With such a value, automatic disconnection may take up to 50 s and clearly this does not meet the requirements for protection against electric shock.

In such circumstances, the installation designer is afforded two further options:

- local supplementary equipotential bonding in accordance with Regulations 413-02-27 and 413-02-28, or
- protection by an RCD.

Where the local supplementary bonding is the chosen option to provide protection against electric shock, consideration should also be given to the thermal effects of the fault current on the distribution cable over the extended period.

Circuit protective conductors   9

9.1 Circuit protective conductors

Where EEBADS is the chosen method for protection against indirect contact, all exposed-conductive-parts are required to be earthed to satisfy Regulation 413-02-06 for TN systems, Regulation 413-02-18 for TT systems and Regulation 413-02-23 for IT systems. This is achieved by connecting exposed-conductive-parts to the MET, via CPCs, and hence to the means of earthing via the earthing conductor. Exposed-conductive-parts may be earthed singly or in groups using a common CPC for a number of circuits.

A CPC therefore provides part of the earth fault loop in the event of an earth fault in the circuit with which it is associated. The main function of a CPC is to carry the earth fault current without damage either to itself or to its surroundings, e.g. insulation. In doing so, the earth fault current can be detected by the protective device for the circuit, be it an overcurrent protective device, such as a circuit-breaker or fuse, or an RCD. The CPC should be efficient and reliable in order for the earth fault current to be detected so that the protective device can provide its function of automatic disconnection within the prescribed time limit.

There are a number of types of CPC in common use, including metal wiring enclosures such as a rigid conduit, trunking, ducting or riser system, as well as the metal covering of cables, such as the sheath or armouring.

There are three instances where a separate CPC is required:

- through a flexible or pliable metal conduit
- subject to specified criteria, in a final circuit intended to supply equipment producing a protective conductor current in excess of 10 mA in normal service
- to connect the earthing terminal of an accessory to metal conduit, trunking or ducting.

Apart from these instances, there are a number of circumstances for which it is recommended that consideration be given to the provision of a separate CPC if metal conduit, trunking or ducting is used, including:

- industrial kitchens, laundries and other 'wet areas'
- locations where chemical attack or corrosion of the metal wiring enclosure or cable sheath may be expected
- circuits exceeding 160 A rating and any location where the integrity of the metal-to-metal joints in the conduit/trunking installation cannot be ensured over the lifetime of the installation. (Where full information concerning correct and reliable earthing is provided by the manufacturer, such systems may be used up to their maximum rating.)

Even where a separate CPC is provided, the metal cable management system is still an exposed-conductive-part containing live conductors and has therefore still to be properly constructed and provide good continuity throughout its run and be earthed.

Where a CPC is used to serve a number of circuits, it should meet the requirement for the most onerous duty. If the CSA of the CPC is determined by selection, then the CSA relating to the larger or largest phase conductor of all the circuits that are served by the common CPC has to be used. However, where the calculation option is used, the most onerous values of the earth fault current and disconnection time should be used. It should be noted that the lower values of the earth fault current may not necessarily produce the lowest I^2t values.

9.2 Cross-sectional areas

The CSA of a CPC is an important consideration because of the high temperatures which can be generated under earth fault conditions in an inadequately sized conductor.

The temperature rise of a protective conductor under earth fault conditions should therefore be limited to acceptable levels so that damage to its insulation and sheath if present or to adjacent material does not occur.

Regulation 543-01-01 calls for the CSA of a protective conductor, including CPCs but excluding bonding conductors, to be not less than that:

- calculated in accordance with Regulation 543-01-03, or
- selected in accordance with Regulation 543-01-04.

Most practitioners will agree that the option of selection is often much easier, although it has to be said that this may produce a larger CSA value for the protective conductor than that obtained by calculation, erring as it does on the safe side.

Regulation 543-01-01 requires that if the determination of the CSA of the phase conductors was undertaken by consideration of the short-circuit current and the earth fault current is expected to be less than the short-circuit current, the option of calculation has to be used.

9.2.1 Calculation of CSA – general case

Turning first to the calculation option, this is achieved by making use of the adiabatic equation given in Regulation 543-01-03, which we initially presented as Equation (3.1) and now repeat here for convenience as (9.1):

$$S = \frac{\sqrt{I^2 t}}{k} \quad (\text{mm}^2) \tag{9.1}$$

where the definitions are as for (3.1).

The value for S obtained from Equation (9.1) represents the minimum CSA value that is required for the CPC. However, there is a minimum value of 2.5 mm² for the CSA of a separate protective conductor that is not part of a cable and is not formed by a wiring enclosure or contained within such an enclosure.

In order to use Equation (9.1), values of I and t have to be determined. The prospective earth fault current I can be calculated using Equation (9.2).

$$I = \frac{U_0}{Z_s} \text{ (A)} \qquad (9.2)$$

where: U_0 is the nominal line voltage to Earth
Z_S is the earth fault loop impedance related to the prescribed disconnection time limit for the circuit

To take an example, consider a 230 V circuit wired in 70 °C PVC cable with an integral copper CPC and protected by a 63 A BS 88-6 fuse. Disconnection is required within 5 s. As given in Table 41D of BS 7671, the maximum Z_s is 0.86 Ω, which gives an earth fault current I of 267 A, as shown in Equation (9.3):

$$I = \frac{U_0}{Z_s} = \frac{230}{0.86} = 267 \text{ A} \qquad (9.3)$$

Having determined I for the corresponding value of disconnection time t we now have to determine a value for k (a constant) by reference to Tables 54B, 54C, 54D, 54E or 54F as appropriate (replicated in Appendix A). As the CPC is incorporated into a cable, Table 54 C is appropriate, from which we obtain a value for k of 115. We can use these values in Equation (9.1), giving a minimum value of the CSA of the CPC of 5.2 mm², as shown in Equation (9.4). Obviously, a 6 mm² CPC would more than meet the minimum CSA obtained by the use of Equation (9.1).

$$S = \frac{\sqrt{I^2 t}}{k} = \frac{\sqrt{267^2 \times 5}}{115} = \frac{597}{115} = 5.2 \text{ mm}^2 \qquad (9.4)$$

There are situations where the use of separate values of I^2 and t is impracticable or unreliable, such as:

▶ where the value of I is so high that a corresponding value of t is not obtainable from the time/current characteristic for the protective device
▶ for short operating times of less than 0.1 s where current asymmetry is significant (e.g. for a protective device installed immediately downstream of the source)
▶ where the protective device is current-limiting and the prospective earth fault current is of such magnitude that the device will limit the current during fault conditions.

In any of the above circumstances, values of $I^2 t$ should be obtained from the energy let-through characteristic for the protective device published by the device manufacturer. The $I^2 t$ value is derived from the manufacturer's characteristic given in a curve giving maximum values of $I^2 t$ as a function of prospective current under stated operating conditions.

9.2.2 Selection of CSA – general case

An easier option for determining the CSA of a CPC is by using Table 54G of BS 7671, replicated here as Table 9.1.

▸ **Table 9.1** Data from Table 54G of BS 7671: minimum CSA of protective conductors in relation to the CSA of associated phase conductors

CSA of phase conductor S (mm²)	Minimum CSA of corresponding protective conductor (mm²)	
	If protective conductor is of the same material as the phase conductor (mm²)	If protective conductor is not of the same material as the phase conductor (mm²)
$S \leq 16$	S	$\frac{k_1}{k_2} \times S$
$16 < S \leq 35$	16	$\frac{k_1}{k_2} \times 16$
$S > 35$	$\frac{S}{2}$	$\frac{k_1}{k_2} \times \frac{S}{2}$

Notes:
1. k_1 is the value for the phase conductor, selected from Table 43A of BS 7671 according to the materials of both conductors and insulation (replicated in Appendix A).
2. k_2 is the value of k for the protective conductor, selected from Tables 54B, 54C, 54D, 54E or 54F of BS 7671, as appropriate (replicated in Appendix A).

Using the selection option to determine the CSA of a protective conductor is fairly straightforward. For a CPC of the same material as the associated phase conductor, reference to Table 9.1 is all that is required, and we find the minimum CSA of the CPC in column 2 of the table. This can be summarised as:

▸ for phase conductors up to and including 16 mm², the minimum CPC is the same CSA as the phase conductor, subject to a minimum of 2.5 mm² or 4 mm² as set out in the following paragraph
▸ for phase conductors above 16 mm² and up to and including 35 mm², the minimum CSA of the CPC is 16 mm²
▸ for phase conductors above 35 mm², the minimum CSA of the CPC is half that of the phase conductor.

For a CPC which is not one of the following, the CSA is required to be not less than 2.5 mm² of copper equivalent if protected against mechanical damage by, for example, a cable sheath, or 4 mm² if not so protected:

▸ an integral part of a cable (such as a 'twin and earth' cable or an armoured cable), or
▸ formed by conduit, ducting or trunking, or
▸ contained in an enclosure formed by a wiring system.

9.2.3 Non-copper CPCs
Where the CPC is not made of copper, an assessment of its copper equivalent is necessary. Equation (9.5) provides the means for this assessment to be made, with the lower limit of CSA of a non-copper CPC being given by $S_{m(min)}$, as required by Regulation 543-01-01:

$$S_{m(min)} = S_{c(min)} \times \frac{k_c}{k_m} \quad (mm^2) \tag{9.5}$$

where: $S_{c(min)}$ is the lower limit given in Regulation 543-01-01 for the CSA of a copper protective conductor (that is, 2.5 or 4 mm², as appropriate),
k_m is the value of k for a protective conductor of the metal other than copper, and
k_c is the value of k for a copper protective conductor.

9.2.4 Evaluation of k

Values of k are obtained from Table 43A and Tables 54B to 54F of BS 7671 which are replicated in Appendix A. Where data is not available from these tables, k may be evaluated by the use of Equation (9.6), which can also be used when the initial temperature differs from that assumed in the tables:

$$k = \sqrt{\frac{Q_c(B+20)}{Q_{20}} \times \ln\left(1 + \frac{\theta_f - \theta_i}{B + \theta_i}\right)} \qquad (9.6)$$

where: Q_c is the volumetric heat capacity of the conductor material,
B is the reciprocal of temperature coefficient of resistivity at 0 °C for the conductor,
Q_{20} is the electrical resistivity of the conductor material at 20 °C,
θ_i is the initial temperature of conductor,
θ_f is the final temperature of conductor, and
ln is log to the base 'e'.

Equation (9.6) can be rewritten as Equation (9.7), and the first part of the expression can be seen to be constant for a particular conductor:

$$k = \sqrt{\frac{Q_c(B+20)}{Q_{20}}} \times \ln\sqrt{\left(1 + \frac{\theta_f - \theta_i}{B + \theta_i}\right)} \qquad (9.7)$$

Table 9.2 provides data for use in Equation (9.7).

Material	B (°C)	Q_c (J/°C mm³)	Q_{20} (Ω mm)	$\frac{Q_c(B+20)}{Q_{20}}$
Copper	234.5	3.45 x 10⁻³	17.241 x 10⁻⁶	226
Aluminium	228.0	2.50 x 10⁻³	28.264 x 10⁻⁶	148
Lead	230.0	1.45 x 10⁻³	214.000 x 10⁻⁶	42
Steel	202.0	3.80 x 10⁻³	138.000 x 10⁻⁶	78

▶ **Table 9.2** Data for use with (9.7) for evaluating k

As an example of the application of the above data and (9.7), take a copper conductor with 90 °C thermosetting with an initial temperature, θ_i, of 90 °C and a final temperature, θ_f, of 250 °C (from Table 43A of BS 7671). Applying Equation (9.7) and using the data given in Table 9.2, we get a value for k of 143, which is in agreement with that given in BS 7671, as confirmed in Equation (9.8):

$$k = 226 \times \sqrt{\ln\left(1 + \frac{\theta_f - \theta_i}{B + \theta_i}\right)} = 226 \times \sqrt{\ln\left(1 + \frac{250 - 90}{234.5 + 90}\right)} = 226 \times \sqrt{\ln(1.493)} = 143 \qquad (9.8)$$

There are situations for which the designer selects a cable larger than would normally be necessary and in excess of that necessary for the particular load, for example because of voltage-drop considerations. In such a case, the initial temperature would be less than that assumed in Table 43A of BS 7671, and consequently a higher value for k would apply. For the purposes of illustrating the point, take a similar cable to that previously described and assume an initial temperature, θ_i, of 50 °C. Using the data of Equation (9.7) we get a value of 165 for k, as confirmed by Equation (9.9):

$$k = 226 \times \sqrt{\ln\left(1 + \frac{250-50}{234.5+50}\right)} = 226 \times \sqrt{\ln(1.703)} = 165 \qquad (9.9)$$

This calculation is only ever likely to be worthwhile where there is a substantial difference in the actual initial temperature to that given in Table 43A of BS 7671.

9.3 Armouring

The armouring of a cable, in addition to providing mechanical robustness and protection against impact, may also be used as a CPC. However, as with all CPCs it is required to meet all the relevant requirements for such conductors. Figure 9.1 shows a typical cross-section of a four-core steel-wire armoured cable with extruded bedding and oversheath.

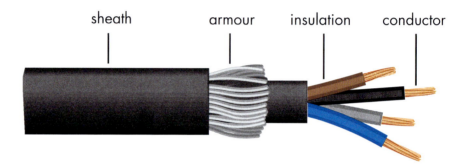

Figure 9.1 Four-core steel-wire armoured cable

The cable armouring, where used as a CPC, will be subjected to thermal stress during an earth fault due to the earth fault current which results. As with any other CPC, the CSA of the armouring is required to be such as to withstand these stresses. In this regard the CSA has to be determined either by calculation or by selection.

9.3.1 Calculation of CSA – armoured cable
For the calculation option, the adiabatic equation has to be used, as given in Equation (9.1). The value for k is obtained from Table 54D of BS 7671.

Having obtained the minimum CSA using Equation (9.1), this has to be checked against the actual CSA of the CPC armouring of the intended cable. Data for this comparison for multi-core cables with copper and solid aluminium conductors to BS 5467 and BS 6724 is to be found in Appendix B.

9.3.2 Selection of CSA – armoured cable
Selection of the CSA of the armouring is much easier than by calculation, and involves the use of the CSA of the phase conductor, and the k values of both the phase conductor and the CPC. However, this method consistently produces a larger minimum CSA requirement than by calculation. Selection should *not* be used where the phase conductor(s) are sized only by consideration of short-circuit currents and where the earth fault current is expected to be less than the short-circuit current.

Data for the selection method is given in Appendix B.

9.3.3 Contribution to earth fault loop impedance

Consideration must be given to the contribution that the armouring makes to the overall earth fault loop impedance, Z_s. There may be instances in which reliance on the armouring as a CPC imposes an unacceptable contribution to the overall impedance to the extent that some other provision for a CPC is required.

9.3.4 Armouring inadequate for a CPC

There will be occasions where the actual CSA of the armouring of a cable is less than the value of S calculated by use of Equation (9.1). To remedy this, the designer has three options to consider:

1 selecting a cable with an extra internal core, and using it as the CPC (note that all the cores are required to be suitably identified – Regulation 514-03-01 refers)
2 selecting a larger cable with a corresponding larger CSA and using the armour as the CPC
3 a separate green-and-yellow covered copper conductor.

For options 1 and 3, it cannot be accurately predicted how the current will divide between what is effectively two parallel conductors (i.e. the core or separate conductor and the armouring) due to the magnetic effect of the armouring. It is therefore important that the additional core or separate CPC is sized as if it alone were to take the earth fault current. In other words, it is not permissible to add the CSAs of the two conductors together.

9.3.5 Termination of armoured cables

It is of paramount importance that armoured cables are terminated in a proper manner and in accordance with the manufacturer's installation instructions. BS 7671 demands the preservation of continuity of protective conductors and specifically requires such conductors to be protected against:

- mechanical damage and vibration
- chemical deterioration and corrosion
- heating effects
- electrodynamic effects (i.e. mechanical forces generally associated with fault currents).

Even where the armouring is not serving as a CPC, it will still need to be earthed as it is an exposed-conductive-part. Armouring is therefore always to be protected against the above detrimental influences and any other stresses which could impair its continuity. Electrical joints such as those between armouring, cable glands and earthing terminals, are required to be soundly made (mechanically and electrically), and, where necessary, suitably protected.

Regulation 8 of the *Electricity at Work Regulations* places an absolute requirement on protective conductor connections to earth and states:

> *a conductor shall be regarded as earthed where it is connected to the general mass of earth by conductors of sufficient strength and current-carrying capability to discharge electrical energy to earth*

As Regulation 8 of the *Electricity at Work Regulations* is an absolute requirement, compliance should be achieved regardless of cost or any other consideration.

It may be questionable whether the termination of steel or aluminium wire armouring with glands into metal gland plates, which themselves may only be bolted to the switchgear or controlgear frame, is adequate. Terminations made in this way will always leave some doubt about the effective current-carrying capacity, which may be of the order of several kiloamperes, through the metalwork joints, some of which may be electrically impaired by paint or other surface finishes. It is therefore always desirable, and often necessary, to employ 'gland tag washers' together with a copper protective conductor with cable lug to bridge the 'gap' between the armouring and the earthing terminal of the item of equipment. Figure 9.2 illustrates an example of a means of terminating an armoured cable. It shows the removable plate at the bottom of the enclosure, together with two retaining screws, through which all the earth fault current would pass were it not for the gland tag washer arrangement.

▶ **Figure 9.2** An example of a means of terminating an armoured cable

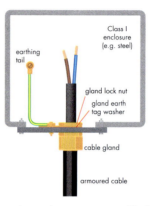

If an armoured cable is to terminate in a non-metallic item of equipment, such as an adaptable box, a gland tag washer will be essential if CPC preservation is to be maintained. The gland tag washer should be clamped between two locknuts to avoid the risk of the degradation of continuity of the connection by creepage of the non-metallic material. Figure 9.3 illustrates such an arrangement.

▶ **Figure 9.3** Termination of an armoured cable at a non-metallic item of equipment

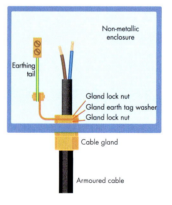

9.4 Steel conduit

Metal conduit (and also steel trunking and ducting) has traditionally been used as a CPC for many years, but recently it has fallen out of favour, with many designers adopting a 'belt and braces' approach by using a separate CPC contained within the wiring system. It has to be said that in many, if not most, cases the additional separate CPC is wholly unnecessary. Additionally, it has to be recognised that a steel containment system is required to be effectively earthed where the contained cables are not sheathed. Thus, it is important that joints in the containment system be mechanically and electrically sound even where a separate CPC is employed.

A metal conduit often has a large enough CSA to allow it to be used as a CPC protective conductor for the circuits whose wiring it contains.

Like other forms of CPC, Regulation 543-01-01 requires the CSA of the conduit to be either:

▸ calculated in accordance with Regulation 543-01-03 (essential where the CSA of the live conductors has been chosen by considering only the short-circuit current and where the earth fault current is predicted to be less than the short-circuit current), or
▸ selected in accordance with Regulation 543-01-04.

Using either option, the CSA of the CPC has to be sufficient so as to limit the temperature rise to acceptable limits, thus avoiding any damage to the conductor insulation and, where applicable, the sheaths of the cables and adjacent surroundings.

9.4.1 Calculation of CSA – steel conduit

As with the calculation of the CSA of any other type of CPC we use Equation (9.1).

To illustrate the calculation option by example, take a 70°C thermoplastic (PVC) insulated copper cable which is to be installed in a steel conduit and the conduit is to be used as the only CPC. The circuit is protected against overcurrent and indirect contact by a 63 A, BS 88-6 fuse. For the purposes of this example, we are given the prospective earth fault current at the furthest point of the circuit as 280 A. As can be seen from Figure 3.3B of Appendix 3 of BS 7671 (replicated here as Figure 9.4), for an earth fault current of this magnitude the overcurrent protective device can be expected to operate in not more than 5 s.

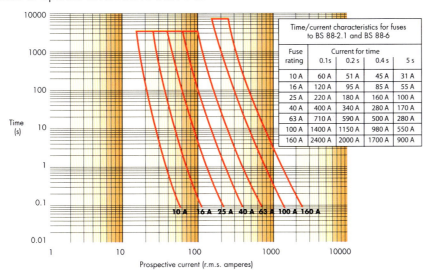

▸ **Figure 9.4**
Time/current characteristics for BS 88-2.1 and BS 88-6 fuses

The other factor required to enable (9.1) to be applied is a value for k, which can be found in Table 54E of BS 7671, and replicated here in Appendix B, and given as 47. Substituting the data into Equation (9.1) we obtain Equation (9.10) which gives a value of S, the minimum CSA for the steel conduit acting as a CPC of:

$$S = \frac{\sqrt{I^2 t}}{k} = \frac{\sqrt{(280^2 \times 5)}}{47} = \frac{\sqrt{(392\,000)}}{47} = \frac{626}{47} = 13.3 \text{ mm}^2 \qquad (9.10)$$

To conclude, where the CSA is calculated, the minimum CSA S of the conduit is required to be not less than 13.3 mm².

Having determined by calculation the minimum CSA, this has to be checked against the CSA data for steel conduit, given in Table 9.3. It will be noted that even the smallest steel conduit has a greater CSA than that required, of 13.3 mm².

Table 9.3 CSAs of heavy-gauge steel conduit from BS 4568: Part 1: 1970

Nominal diameter (mm)	Minimum CSA (mm²)
16	53
20	81
25	103
32	135

9.4.2 Selection of CSA – steel conduit

Selection of the minimum CSA for a CPC provided by a steel conduit embraces the following procedure:

- Utilise the data given in Table 54G of BS 7671, together with that in Tables 43A and 54E of BS 7671 (see Appendix B of this Guide)
- It is then necessary to determine the actual CSA of the conduit (for example, by reference to the product manufacturer or the relevant British Standard, or from Table 9.3 of this Guide)
- Make certain that the actual CSA is not less than the required CSA.

Table 9.4 provides data based on the above procedure. Taking an example of the application of the data, consider a steel conduit enclosing two circuits:

- circuit no. 1: 2.5 mm² with 85 °C thermosetting (rubber) insulation
- circuit no. 2: 6 mm² copper conductors with 70 °C thermoplastic (PVC) insulation.

From the table we can obtain a minimum CSA of 6.2 and 14.7 mm² for the CPC for circuit no. 1 and circuit no. 2, respectively. Comparing the most onerous CSA of 14.7 mm² with data in Table 11.4, we see that any conduit size between 16 and 32 mm will be adequate from this viewpoint.

Table 9.4 Minimum CSA required for a steel conduit, enclosing copper conductors, determined by using formulae from Table 54G of BS 7671

CSA of largest phase conductor (S)	Table 54G formula	Minimum CSA of steel conduit (mm²)			
		Containing 70 °C thermoplastic (PVC) insulated copper cables	Containing 90 °C thermoplastic (PVC) insulated copper cables	Containing 85 °C thermosetting (rubber) insulated copper cables	Containing 90 °C thermosetting insulated copper cables
		$k_1 = 115, k_2 = 47$	$k_1 = 100, k_2 = 44$	$k_1 = 134, k_2 = 54$	$k_1 = 143, k_2 = 58$
1.5	$\frac{k_1}{k_2} \times S$	3.7	3.4	3.7	3.7
2.5		6.1	5.7	6.2	6.2
4		9.8	9.1	9.9	9.9
6		14.7	13.6	14.9	14.8
10		24.5	22.7	24.8	24.7
16		39.1	36.4	39.7	39.4
25	$\frac{k_1}{k_2} \times 16$	39.1	36.4	39.7	39.4
35		39.1	36.4	39.7	39.4
50	$\frac{k_1}{k_2} \times \frac{S}{2}$	61.2	56.8	62	61.6
70		85.6	79.5	86.9	86.3

9.4.3 Maintaining protective conductor continuity

As previously mentioned, whether or not the steel conduit is used as a CPC, the conduit itself is required to be earthed. It is therefore essential that all joints in the conduit are mechanically robust and electrically sound. This requires that the screwed joints in the system, in the conduit couplers and the conduit boxes are clean before being tightened. Similarly, bushes and couplers used for terminating the steel conduit into accessory boxes and the like are required to be tightened by the use of the correct tool. The conduit system is required to be selected with reference to environmental conditions. Black enamelled and galvanised steel conduit is available for use in dry and other conditions, respectively (shown in Figure 9.5). For very arduous duties, stainless steel conduit is also available. Figure 9.6 shows examples of conduit fittings.

▶ **Figure 9.5** Black enamelled and galvanised steel conduit

Galvanised steel conduit boxes

Black enamelled steel conduit boxes

▶ **Figure 9.6** Black enamelled and galvanised steel conduit boxes

9.5 Steel trunking and ducting

The CSA of a CPC formed by steel trunking and ducting is determined in the same manner as for steel conduits, either by calculation using Equation (9.1) or by selection using data given in Table 9.4. The CSAs of typical sizes of steel surface trunking are given in Table 9.5.

▶ **Table 9.5** CSAs of steel trunking to BS 4678: Part 1: 1971

Nominal size (mm x mm)	Minimum CSA (mm²) without lid
38×38	103
50×38	113
50×50	135
75×50	189
75×75	243
100×50	216
100×75	270
100×100	324
150×50	270
150×75	324
150×100	378
150×150	567

9.5.1 Maintaining protective conductor continuity

As with steel conduits, steel trunking will constitute an exposed-conductive-part even where not used as a CPC. It is therefore of paramount importance that sections of trunking, trunking bends and trunking terminations are mechanically robust and electrically sound. Copper links are available to link two sections together, or a section and a bend, although a suitable additional conductor may be required between sections of a metallic wiring system and between such a system and equipment enclosures to which it connects. Trunking bends such as angled bends (e.g. 45°, 90° and 135°) supplied by the manufacturer should always be used. Figure 9.7 illustrates a suitable link intended to bridge the gap between trunking sections.

▶ **Figure 9.7** Trunking sections with a suitable link

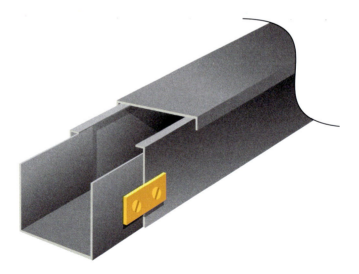

9.6 Metal enclosures

Where a CPC is made up in part by a metal enclosure or frame of low voltage switchgear or controlgear, it is important that effective electrical continuity and fault current carrying capability is maintained, either by construction or by suitable connection.

BS 7671 requires that the CSA of the enclosure or frame must not be less than that calculated by Equation (9.1) or selected by reference to Table 54G of BS 7671. Alternatively, this may be verified by a test procedure in accordance with BS EN 60439-1: *Requirements for type-tested and partially type-tested assemblies.*

If there is doubt as to whether the above requirements are met, a gland earth tag washer and copper protective conductor arrangement should be provided as shown in Figure 9.2. For Class I equipment, the additional gland locknut shown in Figure 9.3 is not essential if the gland earth tag washer can be fitted directly in contact with a cleaned metallic surface of the enclosure. For non-metallic enclosures and Class II equipment the additional locknut should always be used.

9.7 Terminations in accessories

As conductors for safety, the proper termination of CPCs is essential in every instance. This is no less so where the CPC is terminated into an accessory. The effectiveness of the termination is reliant on good workmanship and diligence of the electrician.

For accessories where the CPC is provided by a metal conduit, or by trunking or ducting, or the metal sheath or armour of a cable, Regulation 543-02-07 of BS 7671 requires an earthing tail to be fitted that connects the earthing terminal of the accessory to the earthing terminal incorporated in the associated back box. Figure 9.8 shows this earthing tail fitted so as to link an accessory to a dado-height metallic trunking system and a similar accessory to a flush conduit system back box.

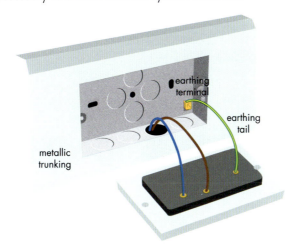

▶ **Figure 9.8** 'Earthing tail' to an accessory

In the example shown in Figure 9.8, the accessory box is earthed to the metal conduit, trunking, ducting (or metal sheath or armour of a cable, if this were to be the wiring system) forming the CPC. The purpose of the earthing tail is not to earth the back box but to earth the accessory by connecting it to the back box and hence to the CPC.

A metal back box for a surface-mounted accessory is an exposed-conductive-part, and a metal back box for a flush-mounted accessory is deemed to be an exposed-conductive-part. Back boxes, like all exposed-conductive-parts, are required to be earthed.

For flush-mounted accessories supplied by a CPC consisting of a single-core cable or a core of a cable connected directly into the earthing terminal of an accessory, the type of the accessory-fixing lugs will dictate whether or not an earthing tail is required.

Flush accessories are often fitted with adjustable accessory-fixing lugs to aid the final levelling of the accessories at second-fix stage. These lugs are of a strap and eyelet arrangement which cannot be relied upon to effectively earth the accessory.

For flush metal back boxes with two fixed accessory-fixing lugs, as shown in Figure 11.9, the box can be considered adequately earthed through the earthing straps and eyelets of the accessory and the fixing lugs on the box. An earthing tail is not therefore essential although for best practice it is highly desirable.

▶ **Figure 9.9** Flush metal box with two fixed accessory-fixing lugs

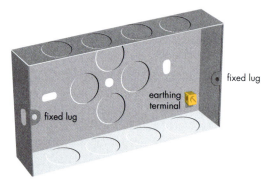

For flush metal back boxes with two adjustable accessory-fixing lugs, as shown in Figure 9.10, an earthing tail is essential because these lugs cannot be relied upon for continuation of the CPC to earth the accessory. Experience has shown that the lugs are subject to corrosion and can present a high-resistance connection which will impede the earth fault current which may prevent the device for automatic disconnection to operate.

▶ **Figure 9.10** Flush metal box with two adjustable accessory-fixing lugs

For flush metal back boxes with one fixed and one adjustable accessory-fixing lug, as shown in Figure 9.11, it is always desirable to provide an earthing tail. However, because some accessories have only one earthing strap and eyelet, it is essential that the earthing eyelet is located at the fixed lug position, otherwise continuation of the CPC will not be maintained and an earthing tail is therefore required to be provided.

▶ **Figure 9.11** Flush metal box with one adjustable and one fixed accessory-fixing lugs

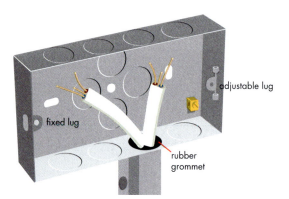

9.8 CPC for protective and functional purposes

Confusion sometimes arises over the difference between protective earthing and functional earthing. As previously mentioned, protective earthing, as the name suggests, is provided for the protection of people, livestock and property. A functional earth is, however, only provided to enable equipment to operate correctly. At no time does the functional earthing offer protection to either the user or the equipment. Examples of the purposes to which functional earthing is put are:

- to provide a 0 V reference point
- to enable an electromagnetic screen to be effective
- to provide a signalling path for some types of communications equipment.

The most common use of functional earthing is for telecommunications purposes. It is permissible for the functional earthing conductor to be terminated onto the electrical installation MET. The wiring used will normally be of copper with a CSA not less than 1.5 mm^2. It should be identified by the colour cream, as recommended in BS 6701: 1994 and Table 51 of BS 7671. Additionally, the functional earthing conductor should have a label (or, alternatively, embossed sheath) reading 'Telecoms Functional Earth' where terminated at the MET. Electrical installation practitioners who come across such functional earthing, which is particularly common in commercial buildings, should not interfere with these connections.

Functional earthing may also be required for other equipment and should be identified by the colour cream. The connection to earth should again be made to the MET and be clearly labelled as to its purpose.

Where earthing arrangements are installed for combined protective and functional purposes, the requirements for protective measures are required always to take precedence. A particular application of this is the use of a combined protective and neutral conductor (PEN) in an installation forming part of a TN-C system. The detailed requirements of BS 7671 for such an installation are tightly drawn. Additionally, an exemption to the *ESQCR 2002* has to be obtained by consumers for adoption of a TN-C system, which is fairly rare.

9.9 Significant protective conductor currents

Most items of current-using equipment exhibit some current flow in their CPC and generally this is of a magnitude that does not cause concern. However, some items are more prone to show evidence of protective conductor currents which cannot be ignored and in these cases provision has to be made to employ CPCs with a higher integrity. This is because if a CPC becomes discontinuous for whatever reason, these protective conductor currents when interrupted will cause a potential difference between items connected to either end of the conductor, and this can present a risk of electric shock. Section 607: *Earthing requirements for the installation of equipment having high protective conductor currents* addresses the particular requirements for the CPCs for such equipment.

Figure 9.12 shows one example of an item of information technology (IT) equipment which is likely to produce some measure of protective conductor current.

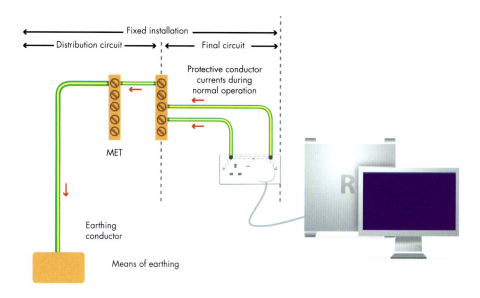

Figure 9.12 Item of IT equipment causing a protective conductor current

In Figure 9.12, the computer is equipped with filters or suppressors which generate a protective conductor current. Other IT equipment, telecommunications equipment and equipment such as fluorescent luminaires are also likely to cause such currents to flow.

Filters are often provided in such equipment to save downstream equipment from harm which might otherwise be caused by transient or switching overvoltages originating from other parts of the fixed installation or from the supply. Where switched mode power supplies are involved, suppressors may serve the purpose of preventing high-frequency noise being transferred into the power supply. Filters and suppressors commonly use capacitive and inductive components between the live conductors (phase and neutral conductors) and the CPC.

As mentioned in Clause 9.8, and as required by Regulation 546-01-01, where the CPC is provided for the combined purpose of protection against indirect contact and for functional purposes, the protective requirements have to take precedence.

In order to improve light output, reduce control gear losses, provide silent operation and reduce perceptible flicker, many fluorescent luminaires operate at high frequency (e.g. 30 kHz). For the purpose of preventing high-frequency noise being impressed on the supply, these luminaires are often fitted with suppressors which produce protective conductor currents. Sufficient quantities of luminaires can produce protective conductor current magnitudes which are significant and of concern.

The risk associated with final circuits with significant protective conductor currents is that resulting from discontinuity of the protective conductor. In installations where there are significant protective conductor currents, serious risks from electric shock can exist from accessible conductive parts connected to protective conductors which are not connected to the MET. This risk is extended to all the items of equipment on the particular circuit, whether or not they individually have significant protective conductor currents. The more equipment that is connected to a circuit, the wider is spread the risk.

IEC Publication 479-1: *The effects of current passing through the human body* (also published as BS PD 6519-1) advises that for currents less than 10 mA passing through the human body, there are usually no harmful physiological effects. As the body currents increase, so the risk of organ damage and probability of ventricular fibrillation increases, as can be seen in Figure 9.13 and Table 9.6.

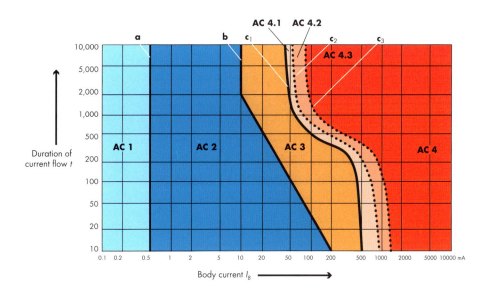

▶ **Figure 9.13** Figure 14 from IEC 479-1

▶ **Table 9.6** The effects of current on the human body (Table 4 of IEC 479-1)

Zone designation	Zone limits	Physiological effects
AC-1	Up to 0.5 mA (line a)	Usually no reaction
AC-2	From 0.5 mA (line a up to line b)	Usually no harmful physiological effects
AC-3	From line b up to curve c_1	Usually no organic damage to be expected. Likelihood of cramp-like muscular contractions and difficulty in breathing for current flow durations longer than 2 s. Reversible disturbances of formation and conduction of impulses in the heart, including atrial fibrillation and transient cardiac arrest without ventricular fibrillation, increasing with current magnitude and time
AC-4	Beyond curve c_1	Increasing with magnitude and time; dangerous pathophysiological effects such as cardiac arrest, breathing arrest and severe burns may occur in addition to the effects of zone 3
AC-4.1	Between c_1 and c_2	Probability of ventricular fibrillation increasing up to about 5 per cent
AC-4.2	Between c_2 and c_3	Probability of ventricular fibrillation up to about 50 per cent
AC-4.3	Beyond curve c_3	Probability of ventricular fibrillation above 50 per cent

Notes:
1 For durations of current flow below 10 ms, the limit for the body current for line b remains constant at a value of 200 mA.
2 With reference to Figure 9.13, as regards ventricular fibrillation, this figure relates to the effects of current which flows in the path left hand to both feet. The threshold values for durations of current flow below 0.2 s apply only to current flowing during the vulnerable period of the cardiac cycle.

When equipment has a protective conductor current of 10 mA, the impedance that allows this at 230 V is 23 000 Ω. The body plus footwear impedance is usually about 2000 Ω (in dry conditions). Consequently, the body impedance does not effectively limit the touch current. If a person touches the exposed-conductive-parts of this equipment when the protective conductor is disconnected, the current conducted through the body is 230 V/(23 000 + 2000 Ω), that is 9.2 mA, a minimal reduction.

From Table 9.6 (Table 4 of IEC 479-1) protective conductor currents exceeding 10 mA can have harmful effects, and precautions need to be taken. The requirements of Section 607 of BS 7671 are intended to increase the reliability of the connection of protective conductors to equipment and to earth, when the CPC current exceeds 10 mA. This normally requires duplication of the protective conductor or an increase in its CSA. The increased CSA is not to allow for thermal effects of the protective conductor currents, which are insignificant, but to provide for greater mechanical robustness and thereby a more reliable connection to earth. Duplication of a protective conductor each with independent terminations is likely to be more effective than an increase in CSA.

9.9.1 Equipment

BS EN 60950, the standard for the safety of IT equipment, including electrical business equipment, requires equipment with a protective conductor current exceeding 3.5 mA to have an internal protective conductor CSA not less than 1 mm² and also requires a label bearing the wording given in Figure 9.14 or Figure 9.15 or similar wording fixed adjacent to the equipment primary power connection.

▶ **Figure 9.14** Equipment warning notice

HIGH LEAKAGE CURRENT
Earth connection essential
before connecting the supply

▶ **Figure 9.15** Equipment warning notice

WARNING HIGH TOUCH CURRENT
Earth connection essential
before connecting the supply

Regulation 607-02-02 requires a single item of equipment, if the protective conductor current exceeds 3.5 mA but does not exceed 10 mA, to be permanently connected or connected by a plug and socket-outlet complying with BS EN 60309-2, as illustrated in Figure 9.16.

▶ **Figure 9.16** BS EN 60309-2 plug [illustration courtesy of Legrand Electric Limited]

If the protective conductor current exceeds 10 mA the requirements of Regulations 607-02-03 and 607-02-04 for a high-integrity protective earth connection should be met.

9.9.2 Labelling at distribution boards

Distribution boards supplying circuits with significant protective conductor currents are required to be labelled accordingly so that persons working on the distribution boards can maintain the protective precautions already in place.

Where a circuit has or is likely to have a significant protective conductor current, the protective conductor connection arrangements at the distribution board will be affected by, for example, the sequence of connections. Additional information, indicating those circuits that have a significant protective conductor current, is required to be made available in the form of a label and should be positioned so as to be visible to a person modifying or extending the circuit, as required by Regulation 607-03-02.

9.9.3 Ring final circuits

Ring final circuits provide duplication of the protective conductor, and if the ends of the protective conductor are separately terminated at the distribution board and at the socket-outlets, the requirements of Section 607 will be met, as shown in Figure 9.17. Socket-outlets are available with two earth terminals for this purpose, both of which should be used to uphold this increased integrity of the CPC.

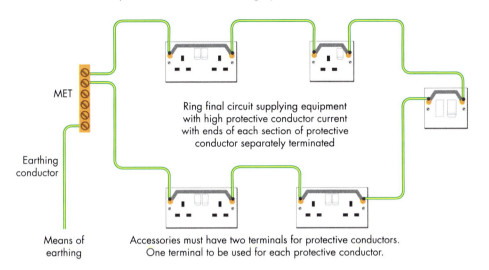

▶ **Figure 9.17** Ring final circuit supplying socket-outlets (total protective conductor current exceeding 10 mA)

9.9.4 Radial final circuits

Radial final circuits supplying socket-outlets where the protective conductor current is expected to exceed 10 mA are required to have a high-integrity protective conductor connection. This can often be most effectively provided by a separate duplicate protective conductor connecting the last socket-outlet directly back to the distribution board as shown in Figure 11.18. This will provide a duplicate connection for each socket-outlet on the circuit. The following requirements apply:

▶ all socket-outlets are required to have two protective conductor terminals, one for each protective conductor
▶ the duplicate protective conductors are required to be separately connected at the distribution board.

To reduce interference effects, the duplicate protective conductor should be run in close proximity to the other conductors of the circuit.

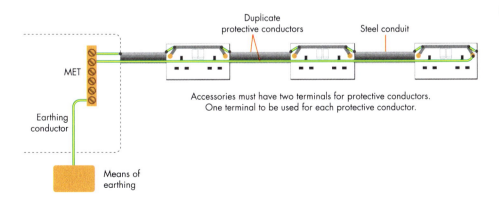

▶ **Figure 9.18** Radial circuit supplying socket-outlets (total protective conductor current exceeding 10 mA), with duplicate protective conductor

Circuit protective conductors
© The Institution of Engineering and Technology

9.9.5 Busbar systems

Busbar systems are often adopted by designers for supplying IT equipment (Figure 9.19). These may be radial 30 or 32 A busbars with tee-offs to individual socket-outlets. The main protective earth (PE) busbar will need to meet one or more of the requirements of Regulation 607-02-04, having a protective conductor with a CSA not less than 10 mm² or duplicate protective conductors each with a CSA sufficient to meet the requirements of Section 543.

▶ **Figure 9.19** Spurs from a 30 or 32 A busbar

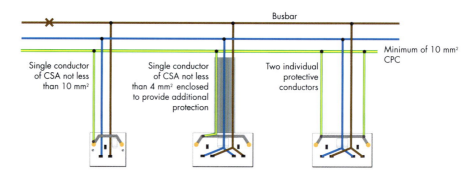

The socket-outlets are often connected as spurs off such a busbar system. If the protective conductor current is expected to be less than 3.5 mA, there is no need for duplication of the protective conductor on the spur from the radial busbar to the socket-outlet. If the protective conductor current is likely to exceed 10 mA then a single copper protective conductor having a CSA of not less than 4 mm² and enclosed to provide additional mechanical stability may be used.

It is necessary for the designer to confirm that the disconnection time of the 30 or 32 A device feeding the busbar is sufficiently fast to provide adiabatic fault protection to the spurred conductors. This almost certainly will be so for, say, 2.5 mm² conductors where a disconnection within 0.4 s is required, but this should be confirmed. Overload protection is not required on the socket-outlet spur from the busbar provided that the spur cable rating exceeds 13 A, since the outlet is a standard 13 A outlet and only plugs fitted with a maximum 13 A fuse can be connected. Hence, overload of the conductor cannot occur, and the 30 or 32 A device can provide adequate fault protection.

9.9.6 Connection of an item of equipment (protective conductor current exceeding 10 mA)

The connection is required to be made by one of three methods:

▶ a permanent connection
▶ a plug and socket-outlet complying with BS EN 60309-2
▶ by employing an earth monitoring system.

Equipment with a protective conductor current exceeding 10 mA should preferably be permanently connected to the fixed wiring of the installation, with the protective conductor meeting the high-integrity requirements detailed below. It is permitted for the final connection between the item of equipment and the wiring of the installation to be made by means of a flexible cable. The high-integrity requirements of Regulation 607-02-04 give five options:

1. A single protective conductor having a CSA of not less than 10 mm².
2. A single copper protective conductor having a CSA of not less than 4 mm², the protective conductor being enclosed to provide additional mechanical protection such as in a flexible conduit or a trunking system.

3 Two individual protective conductors, each one complying with the requirements of Section 543 of BS 7671, which covers the CSA and types of permitted protective conductors. It is permitted for the two protective conductors to be of different types, such as a metallic conduit and a protective conductor enclosed within the same conduit. The two individual protective conductors may also be incorporated in the same multi-core cable provided that the total CSA of all the conductors, including the live conductors, is not less than 10 mm^2. One of the protective conductors may be formed by the metallic sheath, armour or wire braid screen incorporated in the cable (as shown in Figure 9.20), provided that it meets the requirements of Regulation 543-02-05.

4 An earth monitoring system conforming to BS 4444 configured to automatically disconnect the supply to the equipment in the event of a continuity fault in the protective conductor.

5 A double-wound transformer or equivalent in which the input and output circuits are electrically separate. In this case, the CPC is required to connect the exposed-conductive-parts of the equipment and a point of the secondary winding of the transformer to the MET of the installation, and the protective conductor between the equipment and the transformer is required to comply with the requirements of Regulation 607-02-04. It is important to note that the limits on the earth fault loop impedance associated with protection against indirect contact should be met, taking into account the added impedance represented by the transfer function of the transformer.

Equipment with a protective conductor current exceeding 10 mA may be connected to the fixed wiring of an installation by a plug and socket-outlet complying with BS EN 60309-2 (as for equipment with a protective conductor current between 3.5 and 10 mA), provided that one of two additional requirements relating to the CSA of the protective conductor is met. The first requirement is that the CSA of the protective conductor of the associated flexible cable is not less than 2.5 mm^2 for plugs rated at 16 A, and not less than 4 mm^2 for plugs rated above 16 A. Alternatively, the protective conductor of the associated flexible cable should have a CSA that is not less than that of the phase conductor.

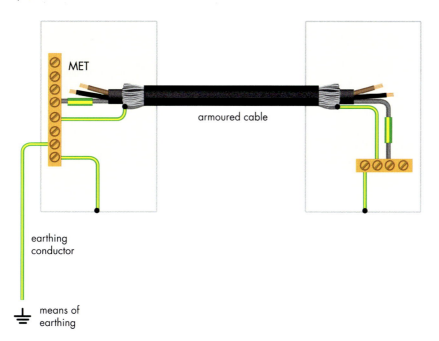

▶ **Figure 9.20** One duplicate protective conductor formed by the armour

Circuit protective conductors

9.9.7 TT systems

If items of equipment having a protective conductor current exceeding 3.5 mA in normal service are to be supplied from an installation forming part of a TT system, the product of the total protective conductor current (in amperes) and twice the resistance of the installation earth electrodes (in ohms) should not exceed 50, as shown in Equation (9.11):

$$2 \times I_{pc} \times R_A \leq 50 \text{ V} \qquad (9.11)$$

where: I_{pc} is the protective conductor current, and
R_A is the resistance to earth of the installation earth electrode.

If this condition cannot be achieved, the equipment should be supplied through a double-wound transformer or equivalent (see Regulation 607-05-01). This is to limit the voltage, a consequence of the protective conductor current, between earth and any equipment connected to the earth electrode, to 25 V.

9.9.8 IT systems

Equipment with a protective conductor current exceeding 3.5 mA in normal use should not be connected directly to an IT system. Connection by a suitable double-wound transformer may be suitable (see Regulation 607-06-01).

9.9.9 RCDs

A protective conductor current in a circuit will be detected by any RCD in that circuit, which could result in the device operating during normal operation or during switching surges.

An RCD should be so selected and the circuits so subdivided that any protective conductor current expected to occur during normal operation of the connected load(s) will be unlikely to cause unwanted operation of the device (Regulation 531-02-04 refers).

As a rule of thumb, the designer should take care that the anticipated total protective conductor current does not exceed 25 per cent of the rated residual operating current, $I_{\Delta n}$, of the RCD. It may be necessary to provide a number of separate RCD-protected circuits, each designed to supply a limited number of items appropriate to their protective conductor current (see Regulation 607-07-01).

9.10 Earth monitoring

Earth proving and earth monitoring systems conforming to BS 4444 are a recognised means of verifying on a continuous basis the integrity of a protective conductor. In certain installations it may be necessary to know that the integrity of a protective conductor has not been adversely affected. In some installations the protective conductor may be particularly vulnerable (for example, trailing cables can suffer mechanical damage or excessive strain).

During the lifetime of an installation, the continuity of a protective conductor may deteriorate due to corrosion, impact or vibration. It may also suffer a loose connection, sustain deterioration or damage or there may be an unintentional disconnection. A protective conductor so affected may then no longer provide an adequate return path for the fault current in the event of an earth fault. Consequently, the effectiveness of protective measures against indirect contact that rely on a sound earth return path may be impaired or even rendered completely ineffective.

Application of this earth proving and protective conductor monitoring might embrace:

- construction sites where flexible cables incorporating protective conductors supply items of moveable plant
- any fixed installations where reassurance of the integrity of the protective conductor is required.
- the supply to equipment with significant protective conductor currents.

Both earth monitoring and protective conductor proving systems require a return path for the proving or monitoring current. The return path generally consists of an insulated conductor which is known as the pilot conductor. Such systems are the subject of BS 4444.

BS 4444: *1989 Guide to Electrical earth monitoring and protective conductor proving* provides definitions:

- Earth proving system: A system providing a means of maintaining a high degree of confidence in the continuity of a protective conductor within an installation.
- Earth monitoring system: A system for providing a means of maintaining a high degree of confidence in the measured impedance of the protective conductor forming part of the earthing arrangements of an electrical system.

A protective conductor proving system verifies the continuity of a selected protective conductor whereas an earth monitoring system monitors the impedance of the selected protective conductor.

A typical earth proving unit contains an extra-low voltage supply denoted by 'V'; a sensing device denoted by 'S'; an alarm denoted by 'A'; and the contacts of a relay or a contactor denoted by 'D', as shown in Figure 9.21. The low/extra-low voltage supply provides a current which circulates in the loop circuit consisting of the protective conductor, a part of the metal casing of the protected Class I equipment, the pilot conductor and the sensing device, S. The sensing device S is energised when sufficient current flows in the loop circuit. Contacts D then close, connecting the load circuit to the supply. If the magnitude of the current falls below a prescribed value, the relay de-energises, disconnecting the load and activating the alarm. In the simple earth proving system shown in the figure, the voltage supply is a transformer; the sensing device, a relay; and the alarm, a warning lamp or audible alarm.

Reference should be made to BS 4444 *Guide to Electrical earth monitoring and protective conductor proving* for further guidance on this aspect of providing high-integrity CPCs.

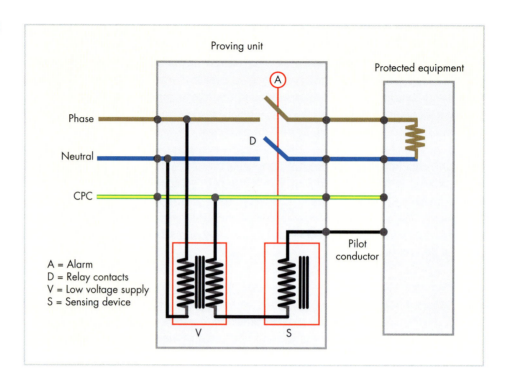

▶ **Figure 9.21** Simplified circuit for earth proving unit

9.11 Proving continuity

Every protective conductor including the earthing conductor and main and supplementary equipotential bonding conductors, as well as CPCs, should be tested to verify that the conductors are electrically sound and correctly connected. Prior to any measurements being taken, BS 7671 requires all protective conductors and their termination to be inspected to verify that they are properly selected and correctly sized, and that their terminations are mechanically and electrically sound.

9.11.1 Radial circuits

In terms of economics, the method described here is the most efficient as not only is the CPC continuity proven but also a value for $(R_1 + R_2)$ is provided. ($(R_1 + R_2)$ is the combined sum of the resistances of the phase conductor and CPC resistance for the circuit.) The measured $(R_1 + R_2)$ resistance can then be added to the upstream earth fault loop impedance to give Z_s, the total earth fault loop impedance. This Z_s value can then be verified to ensure that it is within the prescribed limits according to Tables 41B1, 41B2 or 41D of BS 7671 as appropriate. This method can only be applied to circuits which are 'all insulated'. It cannot be applied to circuits with live conductors enclosed in a steel containment wiring system or for steel-wire armoured cabled circuits which terminate in Class I equipment fixed to structural steel etc. as parallel paths are likely. An ohmmeter capable of measuring low resistance values should be used to make the required measurements.

For compliance with Table 41C of BS 7671, a value for R_2 is required which may be obtained separately by the 'wander lead' method, or by using the measured resistance values of $(R_1 + R_2)$ and the application of Equation (9.12) which assumes that the phase conductor and CPC are made of the same material (e.g. copper) and that they are of the same length. Knowing the precise length, the calculated R_2 can be compared with the data in Table 9.7:

$$R_2 = \frac{R_{1csa}}{R_{1csa} + R_{2csa}} (R_1 + R_2) \ (\Omega) \tag{9.12}$$

where: R_{1csa} is the CSA of the phase conductor
R_{2csa} is the CSA of the CPC

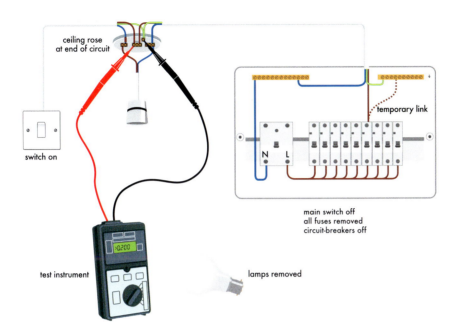

▶ **Figure 9.22** Test method 1 – connections for checking continuity of circuit protective conductors

With reference to Figure 9.22, the procedure for applying test method 1 is:

1 If the test instrument does not include an 'auto-null' facility, or this is not used, the resistance of the test leads should be measured and deducted from the measured resistance readings obtained.
2 Bridge the phase conductor to the CPC at the consumer unit or distribution board so as to include all the circuit.
3 Then measure between phase and earth terminals at each point in the circuit.
4 The measurement taken at the electrically most remote point of the circuit should be recorded on the *Schedule of Test Results* and is the value of $(R_1 + R_2)$ for the circuit under test.

Another test method, test method 2, may be used for checking the continuity of all protective conductors including earthing and bonding conductors.

The procedure for applying test method 2 is:

1 Connect one terminal of the low-resistance test instrument to a long test lead and connect this to the consumer's MET or distribution board earthing terminal.
2 Connect the other terminal of the instrument to another test lead and use this to make contact with the protective conductor at various points on the circuit, such as luminaires, switches, spur outlets etc.
3 The measured resistance of the protective conductor is the value of R_2 and should be recorded on the *Schedule of Test Results* for the circuit under test.

Only measured values of R_2 can be obtained using test method 2.

Table 9.7 provides data for the resistance per metre values for copper and aluminium conductor at temperatures of 20 °C and 70 °C.

▶ **Table 9.7** Values of resistance/metre for copper and aluminium conductors and of $(R_1 + R_2)$/metre in mΩ/m

CSA (mm²)		Resistance/metre or $(R_1 + R_2)$/metre (mΩ/m) at 20 °C		Resistance/metre or $(R_1 + R_2)$/metre (mΩ/m) at 70 °C in compliance with Table 54C	
Phase conductor	Protective conductor	Plain copper	Aluminium	Plain copper	Aluminium
1	-	18.10		21.72	
*1	1	36.20		43.44	
1.5	-	12.10		14.52	
*1.5	1	30.20		36.24	
1.5	1.5	24.20		29.04	
2.5	-	7.41		8.89	
2.5	1	25.51		30.61	
*2.5	1.5	19.51		23.41	
2.5	2.5	14.82		17.78	
4	-	4.61		5.53	
*4	1.5	16.71		20.05	
4	2.5	12.02		14.42	
4	4	9.22		11.06	
6	-	3.08		3.70	
*6	2.5	10.49		12.59	
6	4	7.69		9.23	
6	6	6.16		7.39	
10	-	1.83		2.20	
*10	4	6.44		7.73	
10	6	4.91		5.89	
10	10	3.66		4.39	
16	-	1.15	1.91	1.38	2.29
*16	6	4.23	-	5.89	-
16	10	2.98	-	3.58	-
16	16	2.30	3.82	2.76	4.58
25	-	0.727	1.20	0.87	1.44
25	10	2.557	-	3.07	-
25	16	1.877	-	2.25	-
25	25	1.454	2.40	1.74	2.88
35	-	0.524	0.868	0.63	1.04
35	16	1.674	2.778	2.01	3.33
35	25	1.251	2.068	1.50	2.48
35	35	1.048	1.736	1.26	2.08

* Identifies copper phase/protective conductor arrangement that complies with Table 5 of BS 6004: 1995 for PVC-insulated and sheathed single-, twin- or three-core and CPC cables (i.e. 6241Y, 6242Y or 6243Y cables), and similar cable constructions for thermosetting cables to BS 7211: 1994.

9.11.2 Ring final circuits

Where the continuity of the CPC is being verified together with that of the live conductors, the procedure for proving the continuity of the phase, neutral and protective conductors, and correct wiring of each ring final circuit is:

1. The phase, neutral and protective conductors are identified at the distribution board or consumer unit, and the end-to-end resistance of each is measured separately (Figure 9.23). These resistances are r_1, r_n and r_2 respectively. A finite reading confirms that there is no open circuit on the ring conductors under test. The resistance values obtained should be the same (within 0.05 Ω) where the conductors are the same size. Where the protective conductor has a reduced CSA the resistance r_2 of the protective conductor loop will be proportionally higher than that of the phase and neutral loops e.g. 1.67 times for 2.5/1.5 mm^2 cable. Where these relationships are not achieved then either the conductors are incorrectly identified or there is something wrong at one or more of the accessories.

2. The phase and neutral conductors are then connected together so that the outgoing phase conductor is connected to the returning neutral conductor and vice versa (Figure 9.24). The resistance between the phase and neutral conductors is measured at each socket-outlet. The readings taken at each of the socket-outlets wired into the ring will be substantially the same and the value will be approximately one-quarter of the resistance of the phase plus the neutral loop resistances, i.e. $(r_1 + r_n)/4$. Any socket-outlets wired as spurs will have a higher resistance value due to the resistance of the spur conductors.

   Note that where single-core cables are used, readings should be taken to verify that the phase and neutral conductors at opposite ends of the ring final circuit are connected together. An error in this respect will be apparent from the measured results taken at the socket-outlets. The measurements will progressively increase in value as readings are taken towards the midpoint of the circuit, then decrease again towards the other end of the circuit.

3. Step 2 is then repeated, this time with the phase and CPC cross-connected, as shown in Figure 9.25. The resistance between the phase and CPC is measured at each of the socket-outlets. The measurements obtained at each of the socket-outlets wired in the ring will be substantially the same and the value will be approximately one-quarter of the phase plus the CPC loop resistances, i.e. $(r_1 + r_n)/4$. As before, a higher resistance value will be measured at any socket-outlet wired as a spur. The highest measured value represents the maximum $(R_1 + R_2)$ of the circuit and should be recorded on the *Schedule of Test Results*. This value can be used to determine the earth fault loop impedance, Z_s, of the circuit to verify compliance with the prescribed limits given in BS 7671.

   The sequence of test also verifies the polarity at each socket-outlet and other accessory unless the tests have been carried out on the terminal on the reverse side of the socket-outlet. In such cases, a visual inspection is required to confirm correct polarity connections at the socket-outlet and other accessory such as a connection unit, and dispenses with the need for a separate polarity test.

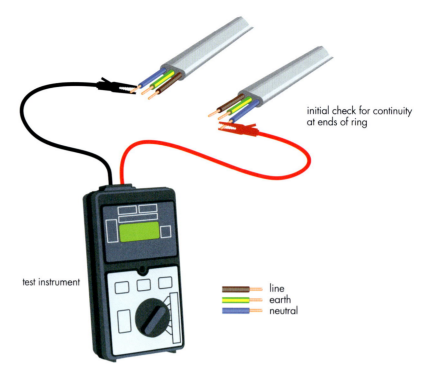

▶ **Figure 9.23** Initial check of continuity at ends of the circuit protective conductor

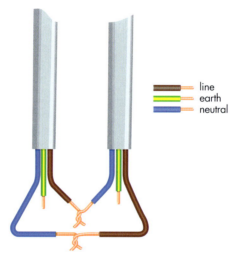

▶ **Figure 9.24** Connections of phase and neutral conductors

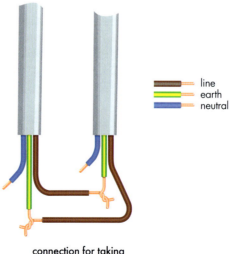

▶ **Figure 9.25** Connections of phase and CPC

Particular issues of earthing and bonding

10.1 Clean earths

A 'clean earth' is a particular form of functional earthing and is defined in BS 7671 as a connection to Earth necessary for proper functioning of electrical equipment. A clean earth is more accurately described as a low-noise earth in which the level of conducted or induced interference from external sources does not produce an unacceptable incidence of malfunction in the data-processing or similar equipment to which it is connected.

Where IT equipment is under consideration, functional earthing and a clean earth or low-noise earth should also be considered.

The protective conductors in a building are subject to transient voltages relative to the general mass of Earth. These transient voltages are termed 'earth noise' and are often caused by load switching. They may also be generated by the charging of an equipment frame via the stray capacitance from a low voltage circuit, or mains-borne transients may be coupled into the earth conductor or frame from supply conductors.

As earth noise can cause malfunction, manufacturers of large computer systems usually make specific recommendations for the provision of a 'clean' mains supply and a 'clean' earth. The equipment manufacturer's guidance should be taken for such installations.

A dedicated earthing conductor may be used for a computer system, provided that:

1 All accessible exposed-conductive-parts of the computer system are earthed, the computer system being treated as an 'installation' where applicable.
2 The MET or bar of the computer system (installation) is connected directly to the building MET by a protective conductor (see Regulation 413-02-03); extraneous-conductive-parts within reach of the computer systems are bonded, but not via the protective conductor referred to in 1 above (see Regulation 542-04-01).

Supplementary bonding between extraneous-conductive-parts and the accessible conductive parts of the computer system is not necessary.

Functional earthing conductors used in providing a clean earth are required to be identified by the colour cream, according to Table 51 of BS 7671.

10.2 PME for caravan parks

Regulation 9(4) of *ESQCR 2002* prohibits the connection of any metalwork in a caravan (or boat) to the combined protective and neutral conductor of the public electricity network. Therefore, the earthing contacts and extraneous-conductive-parts in the form, for example, of metal cases of caravan pitch socket-outlets should not be connected to a PME earthing terminal, where one is made available by the electricity distributor at the site supply intake position.

Figure 10.1 shows a typical caravan park distribution layout where the park supply distribution is connected to the incoming supply with PME. The park installation consists of distribution circuits to the two pillars which then feed the individual caravans.

The park installation is therefore part of a TN-C-S system which is permitted by BS 7671. However, Regulation 608-13-05 requires that the socket-outlets on the pillar are protected by a residual current device with a rated residual current, $I_{\Delta n}$, of not more than 30 mA and that their protective conductors should not be connected to the PME terminal. Instead they must be connected to an earth electrode separate from the PME. The installation downstream of the socket-outlet is therefore part of a TT system.

▶ **Figure 10.1** A typical caravan park distribution arrangement

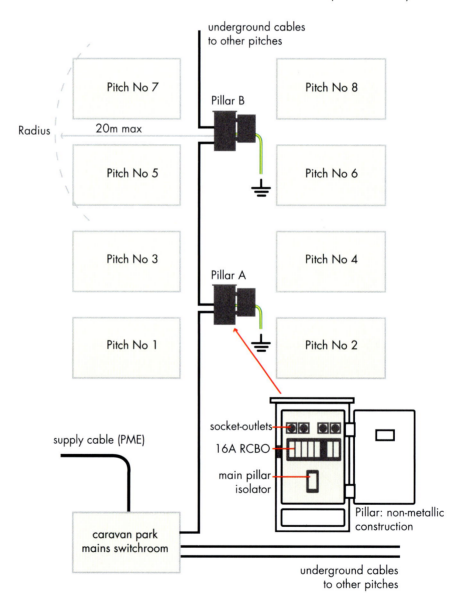

Instead, as required by Regulation 608-13-05 in such circumstances, the protective conductor of each socket-outlet circuit should be connected to an installation earth electrode and the requirements for a TT system should be met.

The demarcation of the park pitch TT system from the electricity distributor's PME earthing terminal may be made at one of a number of places such as:

▶ at the pitch supply position, as shown in Figure 10.2
▶ at the consumer's distribution position, as show in Figure 10.3.

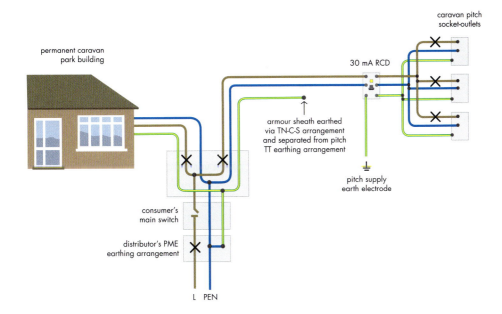

▶ **Figure 10.2**
Demarcation of TN-C-S/TT systems at a park pillar

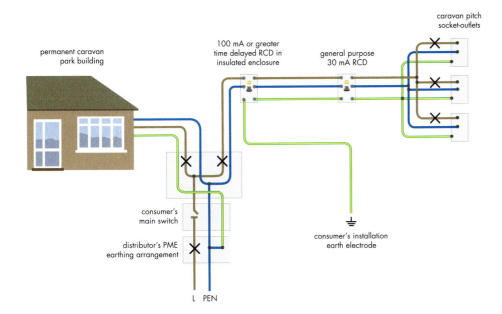

▶ **Figure 10.3**
Demarcation of TN-C-S/TT systems at the consumer's intake position

10.3 Exterior semi-concealed gas meters

The installation of outdoor semi-concealed gas meter boxes has been common practice for a number of years.

Regulation 547-02-02 requires the connection of a main equipotential bonding conductor to be made as near as practicable to the point of entry of a gas service into the premises to the consumer's hard metal pipework before any branch pipework. Where practicable, this connection is required to be made within 600 mm of the meter outlet union or at the point of entry to the building where the meter is external. Although it is generally preferable for a suitable bonding connection to be made inside the premises, this may not always be practicable.

In many situations it is not possible to connect the main equipotential bonding conductor at the point that the gas installation pipe enters the building. In such cases, it is not desirable to make the bonding connection between the gas meter box and where the gas installation pipe enters the building outside because of the likelihood of corrosion and/or mechanical damage.

An alternative and most practical solution is to make the bonding connection inside the meter box itself. The gas distributor will normally provide an earth tag washer on the meter outlet adaptor for this purpose. In cases where such a washer has not been fitted, liaison with Transco will be required.

The bonding conductor should be passed through the pre-drilled aperture in the right-hand side of the meter box and attached using a crimped lug connection to the earth-tag washer. As required by Regulation 514-13-01 a warning notice reading 'SAFETY ELECTRICAL CONNECTION – DO NOT REMOVE' should also be attached. It is essential that the bonding conductor and its connection do not interfere with the integrity of the meter box.

The *Gas Safety (Installation and Use) Regulations 1994* (as amended) prohibit a bonding conductor passing through the same hole as a gas pipe. The bonding conductor external to the box is required to pass through a separate hole in the external wall above the damp-proof course, and should be as short as possible to minimise possible mechanical damage.

The hole for the bonding conductor should be effectively sealed on both sides. The bonding conductor should then be run to the MET of the installation, where it should be arranged or marked so that it can be identified for inspection, testing, repair or alteration of the installation in accordance with Regulation 514-01-02. Figure 10.4 illustrates the technique.

▶ **Figure 10.4** Outdoor gas meter box

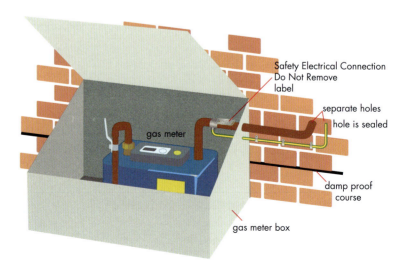

10.4 Small-scale embedded generators

In recent times the use of small-scale embedded generators (SSEGs) has become increasingly commonplace and deserves a mention in this Guidance Note.

10.4.1 Statutory regulations

To set the scene, *ESQCR 2002* exempts sources of energy with an electrical output not exceeding 16 A per phase delivered at low voltage (230 V) from Regulations 22(1)b and 22(1)d, and such sources of energy are addressed here in some detail to explain the earthing requirements.

The requirements for such small generators addressed in Regulation 22(2) of *ESQCR 2002* are:

- (b) the source of energy is configured to disconnect itself electrically from the parallel connection where the distributor's equipment disconnects the supply of electricity to the consumer's installation, and
- (c) the person installing the equipment ensures that the distributor is advised of the intention to use the source of energy in parallel with the network before or at the time of commissioning the source.

Regulation 22(1) requires that 'no person shall install or operate a source of energy which may be connected in parallel with a distributor's network unless he:

- (a) has the necessary and appropriate plant and equipment to prevent danger or interference with that network or with the supply to other consumers so far as reasonably practicable
- (c) where the source of energy is part of a low voltage consumer's installation, complies with the provisions of the British Standard Requirements'.

The British Standard Requirements referred to in Regulation 22(1) are BS 7671 *Requirements for electrical installations (IEE Wiring Regulations)*, and Section 551: *Generating sets* details the particular requirements relating to generators in general.

SSEGs, similar to other electrical equipment, should be type-tested and approved by a recognised body.

10.4.2 Engineering Recommendation G83

To assist network operators and installers, the Energy Networks Association has prepared Engineering Recommendation G83: *Recommendations for the connection of small-scale embedded generators (up to 16 A per phase) in parallel with public low voltage distribution networks.* The guidance given here is intended to replicate the requirements of this Engineering Recommendation as they would apply to persons responsible for electrically connecting SSEGs.

The Engineering Recommendation is for all SSEG installations with an output up to 16 A including:

- domestic combined heat and power
- hydro
- wind power
- photovoltaic
- fuel cells.

Engineering Recommendation G83 incorporates forms, which define the information which is required by a public distribution network operator for a SSEG which is connected in parallel with a public low voltage distribution network. Supply of information in this form, for a suitably type-tested unit, is intended to satisfy the legal requirements of the distribution network operator and hence will satisfy the legal requirements of *ESQCR 2002*.

The installation that connects the embedded generator to the supply terminals is required to comply with BS 7671.

A suitably rated overcurrent protective device is required to protect the wiring between the electricity supply terminals and the embedded generator. The SSEG should be connected directly to a local isolating switch. For single-phase machines, the phase and neutral should be isolated and for multi-phase machines all phases and neutral should be isolated. In all instances the switch, which is required to be manually operated, should be capable of being secured in the 'OFF' isolating position. The switch should be located in an accessible position in the customer's installation. See Figure 10.5.

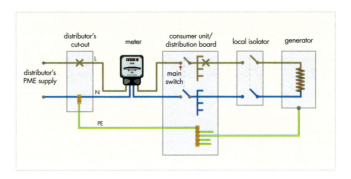

▶ **Figure 10.5** Isolation of a SSEG

Regarding isolation of SSEGs, Regulation 551-07-04 of BS 7671 states:

> *Means shall be provided to enable the generating set to be isolated from the public supply. The means of isolation shall be accessible to the distributor at all times.*

10.4.3 Means of isolation

The means of isolation from the public supply is likely to have to be an additional switch to the generator isolator previously discussed. It is believed that electricity distribution network operators are prepared to accept the distribution cut-out (fused unit) as the means of isolation.

Where a SSEG is operating in parallel with a distributor's network, there should be no direct connection between the generator winding (or pole of the primary energy source in the case of the photovoltaic array or fuel cell) and the network operator's earth terminal (Figure 10.6). All earthing arrangements are required to comply with BS 7671.

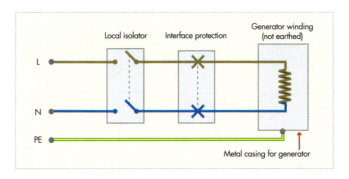

▶ **Figure 10.6** Earthing for parallel operation of a SSEG

10.4.4 Warning notices

Warning notices to indicate the presence of the SSEG within the premises will be required at:

- the supply terminals (fused cut-out)
- the meter position
- the consumer unit
- all the points of isolation.

Health and Safety (Safety Signs and Signals) Regulations 1996 stipulate that the labels should display the prescribed triangular shape and size using black on yellow colours. A typical label both for size and content is shown below in Figure 10.7.

▶ **Figure 10.7** Warning notice – dual supplies

10.4.5 Up-to-date information

Additionally, the Engineering Recommendation requires up-to-date information to be displayed at the point of connection with a distributor's network as follows:

- A circuit diagram showing the relationship between the embedded generator and the network operator's fused cut-out. This diagram is also required to show by whom the generator is owned and maintained.
- A summary of the protection's separate settings incorporated within the equipment.

Figure 12.8 is an example of the type of circuit diagram that is required to be displayed. This diagram is purely for illustrative purposes and not intended to be fully descriptive.

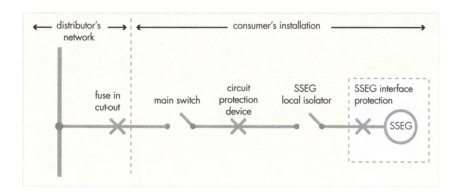

▶ **Figure 10.8** An example of circuit diagram for a SSEG installation

The installer is required to advise that it is the user's responsibility to ensure that this safety information is kept up to date. The operating and maintenance instructions for the installation are required to contain the manufacturer's contact details, e.g. name, telephone number and web address.

Annex B of Engineering Recommendation G83 specifies the particular requirements for combined heating and power sets. These are likely to be the most common type of set encountered by the electricity installer. They can be incorporated into a household gas boiler to generate electricity.

10.4.6 The Stirling engine
Most small combined heat and power generators embody a Stirling engine (Figure 10.9).

▶ **Figure 10.9** The Stirling engine

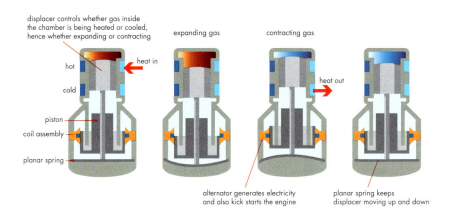

The Stirling engine does not burn the gas within the cylinder as in an internal combustion engine. The power to the engine is delivered by the combustion gases from the gas burner and the energy transfer is in effect achieved by the temperature difference between the burner exhaust gases and burner input air or circulating water.

The basic principle is that the sealed gas within the engine is heated by the burner gases and expands. On expansion the driven piston generates electricity in the winding and compresses the planar spring. In the expanded, say, down position of the piston the gas within the piston is subject to the cooling effect of the cooler input air (or circulated water) and contracts, assisted by the planar spring. At the compression position the gas in the cylinder is now heated by the burner gases only and is not subject to the cooling effects of the input air.

10.4.7 Guidance Note 7: *Special Locations*
Chapter 12 of IEE Guidance Note 7: *Special Locations* reproduces the requirement in the draft IEC Standard for solar photovoltaic power systems. The requirements are for power photovoltaic systems that will generate only when run in parallel with the electricity supply. Requirements for photovoltaic power supply systems which are intended for stand-alone operation are under consideration by the IEC and are not considered here.

10.5 Mobile and transportable units

Chapter 17 of IEE Guidance Note 7: *Special Locations* addresses mobile and transportable units and is based on the draft CENELEC and International Electrotechnical Commission Standard IEC 364-7-717/Ed1: *Requirements for special installations or locations – Mobile or transportable units*. Additional guidance is offered to cover potential UK user requirements.

10.5.1 The term 'mobile or transportable unit'
The term 'mobile or transportable unit' is intended to include a vehicle and/or transportable structure in which all or part of a low voltage electrical installation is contained.

'Units' may either be 'mobile or transportable' – self-propelled or towed vehicles – or 'transportable' – cabins or transportable containers placed *in situ* by other means.

10.5.2 Examples of mobile or transportable units

Examples of units within this scope are broadcasting vehicles, ambulances, fire-engines, mobile workshops, military units and construction site cabins.

Exclusions include:

- transportable generating sets
- marinas and pleasure craft
- mobile machinery
- caravans and other leisure accommodation vehicles
- traction equipment of electric vehicles.

10.5.3 The risks

The risks associated with mobile and transportable units arise from:

- Risk of loss of connection to earth, due to use of temporary cable connections and long supply cable runs; the repeated use of cable connectors which may give rise to 'wear and tear' and the potential for mechanical damage to these parts.
- Risks arising from the connection to different national and local electricity distribution networks, where unfamiliar supply characteristics and earthing arrangements are found.
- Impracticality of establishing an equipotential zone external to the unit.
- Open-circuit faults of the PEN conductor of PME supplies raising the potential of all metalwork (including that of the unit) to dangerous levels.
- Risk of shock arising from significant functional currents flowing in protective conductors – usually where the unit contains substantial amounts of electronics or communications equipment.
- Vibration while the vehicle or trailer is in motion, or while a transportable unit is being moved – causing faults within the installation in the unit.

10.5.4 Reduction of risks

Particular requirements to reduce these risks include:

- Checking the suitability of the electricity supply before connecting the unit.
- Installing an additional earth electrode where appropriate.
- A regime of regular inspection and testing of connecting cables and their couplers, supported by a log-book system of record keeping.
- Recommended use of stranded or flexible cables with a CSA of 1.5 mm^2 or greater for internal wiring with the provision of additional cable supports and stranded conductors.
- Recommended use of stranded or flexible cables with a CSA of 2.5 mm^2 or greater for cables supplying the power to the mobile units.
- Protection of users of equipment outside the unit by the use of 30 mA RCDs.
- The use of RCDs to provide protection against indirect contact.
- The use of electrical separation, by means of either an isolating transformer or an on-board generator.
- The use of earth-free local equipotential bonding, where practicable.
- Selective use of Class II enclosures.
- Clear and unambiguous labelling of units, indicating types of supply which may be connected.
- Particular attention paid to the maintenance and periodic inspection of installations.

10.5.5 Supplies

With regard to supplies, the use of a TN-C system is not permitted inside any unit. Regulation 8(4) of the *ESQCR 2002* forbids the use of combined neutral and protective conductors in a consumer's installation.

The following methods of electricity supply to a unit are acceptable:

1 Connection to a low voltage generating set in accordance with Section 551 of BS 7671.

2 Connection to a fixed electrical installation in which the protective measures are effective; that is, to a TN or TT earthing system.
- Protection against indirect contact by automatic disconnection of supply using an RCD
- Protection by earth-free local equipotential bonding
- Protection against indirect contact by RCD with the use of transformers to provide flexibility for single- and three-phase connection arrangements.

3 Connection through means such as a double-wound transformer providing simple separation from a fixed electrical installation:
- with an internal IT system and an earth electrode
- with an internal IT system, in conjunction with an insulation monitoring device and automatic disconnection of supply after first fault, with or without an earth electrode
- with an internal TN system, with or without an earth electrode.

4 Connection through means (safety isolating transformer) providing electrical separation from a fixed electrical installation.

10.5.6 TN-C-S with PME

Regulation 9(4) of *ESQCR 2002* forbids a distributor's combined neutral and protective (PEN) conductor (from a PME network) being directly connected to the extraneous-conductive-parts of a caravan. For safety reasons the DTI advise that the same prohibition should apply to mobile and transportable units.

Regulation 9(4) of the *ESQCR* relates to the distributor's networks only. In most cases mobile and transportable units, such as outside broadcast vehicles, would be connected to consumers' installations, in which case the governing legislation would be *the Electricity at Work Regulations 1989*, in particular Regulations 8 and 9.

These two regulations require precautions to be taken to prevent danger arising as a result of a fault in the distributor's network, in particular networks with a combined neutral and protective (PEN) conductor; that is PME networks. A particular precaution required by BS 7671 is main bonding to all incoming services. Another precaution that can be taken is the connection of earth electrodes to the MET, e.g. underground structural steelwork, water pipes, earth rods or plates. This may be achieved as a consequence of main bonding or by installing additional earth rods and/or tapes. Where mobile units are manned by electrically competent persons, then they may be required to confirm the adequacy of the earthing of the installation to which the unit is to be connected. Otherwise the site will need to be checked in advance. Where an electricity supply is provided solely for the use of mobile units, say at a pole-mounted box at a showground or racecourse, earth electrodes will need to be installed as permanent features of the supply.

10.5.7 Protection against direct contact

Protection by placing out of reach is not a permitted protective measure.

Additional protection against direct contact by the use of RCDs with a rated residual operating current, $I_{\Delta n}$, not exceeding 30 mA is necessary for all socket-outlets intended to supply current-using equipment outside the unit. This requirement does not apply to socket-outlets which are supplied from circuits protected by SELV, PELV or electrical separation.

10.5.8 Protection against indirect contact

Chapter 17 of IEE Guidance Note 7: *Special Locations* gives comprehensive guidance on the conditions for the various connections depending on system type and the adopted protective measures against indirect contact.

10.6 Highway power supplies and street-located equipment

By definition, highway power supplies include the complete highway installation comprising distribution boards, final circuits and street-located equipment.

Part 2 of BS 7671 provides the following definitions:

- *Highway*. A highway means any way (other than a waterway) over which there is public passage and includes the highway verge and any bridge over which, or tunnel through which, the highway passes.
- *Highway distribution board*. A fixed structure or underground chamber, located on a highway, used as a distribution point, for connecting more than one highway distribution circuit to a common origin. Street furniture which supplies more than one circuit is defined as a highway distribution board. The connection of a single temporary load to an item of street furniture shall not in itself make that item of street furniture into a highway distribution board.
- *Highway distribution circuit*. A band II circuit connecting the origin of the installation to a remote highway distribution board or items of street furniture. It may also connect a highway distribution board to street furniture.
- *Highway power supply*. An electrical installation comprising an assembly of associated highway distribution circuits, highway distribution boards and street furniture, supplied from a common origin.

Section 611 of BS 7671 sets out the particular requirements for such locations, which complement or modify the general requirements elsewhere in the standard. Chapter 11 of IEE Guidance Note 7 provides comprehensive guidance as to the application of Section 611 which applies to street furniture installations on the public highway or on private land such as car parks, public parks and private roads. The guidance given here only concerns issues related to earthing and bonding specific to such locations.

10.6.1 Street furniture and street-located equipment

Typical examples of street furniture are:

- road lighting columns
- traffic signs
- footpath lighting
- traffic control and surveillance equipment.

Typical examples of street-located equipment are:

- bus shelters
- telephone kiosks
- car park ticket dispensers
- advertising signs.

The above list of examples is not exhaustive.

10.6.2 Street furniture access doors

Doors providing access to electrical equipment contained in street furniture provide a measure of protection against interference. However, the likelihood of removal or breakage is such that a door less than 2.5 m above ground level cannot be relied upon to provide protection against direct contact. Therefore, equipment or barriers within the street furniture are required to prevent contact with live parts by a finger (IP2X or IPXXB according to BS EN 60529).

10.6.3 Earthing of street furniture access doors

A street-located equipment door which has no electrical equipment mounted upon it and is not likely to be in contact with wiring is neither an exposed-conductive-part nor an extraneous-conductive-part. Whilst the street lighting column or furniture frame may be an exposed-conductive-part, the door is not, and therefore does not need to be earthed.

10.6.4 Earthing of Class I equipment within street furniture and street-located equipment

Where protection against electric shock is by EEBADS, Class I equipment within an item of street furniture or street-located equipment must be earthed. Similarly, where the street furniture or street-located equipment forms an enclosure for non-sheathed cables, the enclosure must be earthed.

10.6.5 Distribution circuits

Items of fixed equipment supplied from a highway distribution circuit are generally in contact with the ground, and they are no more likely to be subject to bodily contact during a fault condition than items of fixed equipment within a building. BS 7671 already recognises a general 5 s disconnection time for fixed equipment. Present-day practice in highway power supplies does not indicate any need to have a shorter disconnection time. Generally, a disconnection time of 5 s is achievable on the circuit up to the controlgear. Where the fault occurs downstream of the controlgear, the ballast would reduce the energy of the fault and the touch voltage accordingly.

10.6.6 Temporary supplies

Before connecting any additional load the available supply should be accessed and deemed suitable to accept the increase in load safely.

It is important to protect against damage to the existing permanent equipment in the base compartment of an item of street furniture when connecting up a temporary installation. The direct connection of temporary loads to existing fixed installations can lead to damage of the cores of permanent wiring cables. Where it is envisaged that temporary connections will be re-used in the future, it is considered preferable to fit a socket-outlet to feed the temporary load, which can then be left in position after the load has been disconnected and re-used at a later date.

Also, it is intended that access to the base compartment of street furniture and street-located equipment by unauthorised persons should not be made significantly easier by the addition of a temporary supply unit. Whatever the type of unit used, it should be capable of being secured in place by, for example, stainless steel banding and is required to totally cover the base compartment aperture. It is recommended that a socket-outlet be mounted inside these units, preferably behind a lockable access door, to facilitate connection to and disconnection from the unit of the temporary load, without having to remove the unit itself. It is important to note that whilst other equipment is permitted to have a 5 s disconnection time, socket-outlets are restricted to 0.4 s.

It is recommended that temporary supply units which are designed to fit externally over the base compartment aperture of street furniture are manufactured from corrosion-resistant material and fitted with suitable seals to prevent ingress of water.

Temporary power supplies are required to be generally in accordance with the requirements for construction sites. This can be considered to imply a requirement for reduced disconnection times, and may require the installation of a residual current device or a 110 V centre-tapped to earth transformer (CTE), as detailed in Figure 10.10.

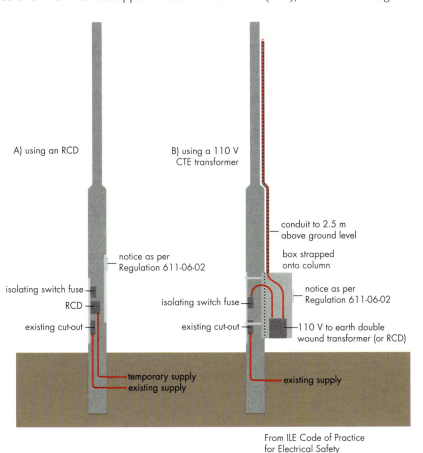

▶ **Figure 10.10**
Temporary supplies from columns – alternative arrangements
a Using an RCD
b Using a 110 V CTE transformer

10.7 Suspended ceilings

Parts 1 and 3 of the British Standard for suspended ceilings, BS 8290: 1991, contain a number of design and maintenance requirements relating to the earthing and bonding of conductive parts of ceiling suspension systems with which, in practice, it would be unworkable to comply. The standard, now withdrawn, caused some uncertainty for the electrical installation designer about the earthing and bonding requirements as they relate to suspended ceilings.

Fortunately for the electrical designer, those electrical safety requirements were applicable only where the conductive parts of the ceiling were going to be used to conduct an earth fault current, or to act as an equipotential bonding conductor. In practice, the conductive parts of a ceiling suspension system cannot be correctly described as exposed-conductive-parts, nor extraneous-conductive-parts. Consequently, earthing or equipotential bonding of such ceiling suspension systems need not receive further consideration.

The exposed-conductive-parts of Class I equipment are required to be connected to the MET of the installation by a protective conductor designed to conduct any earth fault current. Class II equipment is designed such that any basic insulation fault in the equipment cannot result in a fault current flowing into any conductive parts with which it may be in contact. The conductive parts of a suspended ceiling incorporating Class I and/or Class II equipment are not therefore intended to conduct an earth fault current, and so such parts need not be intentionally earthed. Some conductive parts of a suspended ceiling may be earthed, however, by virtue of fortuitous contact with the exposed-conductive-parts of Class I equipment.

The conductive parts of a suspended ceiling will not introduce a potential that does not already exist in the space in which the ceiling is installed. In normal circumstances therefore, there is no need to arrange for the conductive parts of the ceiling to be equipotentially bonded, which would be unnecessary as well as difficult and costly to achieve.

The installation of all electrical equipment, including wiring systems above and incorporated in a suspended ceiling, should fully comply with the requirements of BS 7671 if the risk of electric shock from the ceiling is to be avoided. In particular, cables for fixed wiring should be supported continuously or at appropriate intervals, independently of the ceiling. The method of support is required to be such that no damage or undue strain occurs to the conductors, their insulation or terminations. Figure 12.11 shows a typical arrangement of supporting cables above a suspended ceiling.

▶ **Figure 10.11** Ceiling cables suspended on a catenary

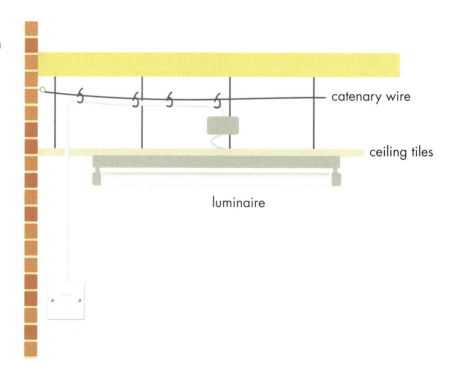

10.8 Exhibitions, shows and stands

Chapter 13 of IEE Guidance Note 7 provides advice on electrical installations at exhibitions, shows and stands, from which the guidance given here concerning earthing and bonding and related matters is taken.

Protective measures against indirect contact by non-conducting location and by earth-free equipotential bonding should not be used.

10.8.1 Protection by automatic disconnection of supply
Because of the practical difficulties of carrying out equipotential bonding to all accessible extraneous-conductive-parts a TN-C-S system is not appropriate for temporary and/or outdoor installations.

A TN-S system would be acceptable where such a supply was available from the distributor. It is most likely and often preferable for TT systems to be adopted. IEC Standard 60364-7-11 disallows both TN-C and TN-C-S systems.

10.8.2 Distribution circuits
Where there is increased risk of damage to cables, disconnection of the supply should be provided by an RCD with a rated residual operating current not exceeding 500 mA. To provide for discrimination with RCDs protecting final circuits, the RCD should be type S to BS EN 61008 or BS EN 61009 or time-delayed to BS 4293.

The installation of RCDs will also increase the protection against the risk of fire arising from leakage currents to earth.

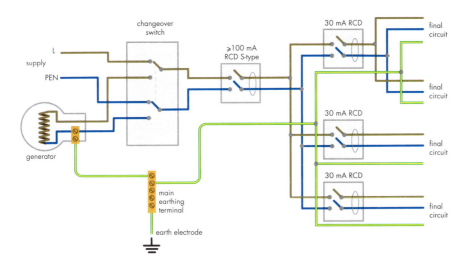

▶ **Figure 10.12**
Exhibition/show distribution layout with standby generator

10.8.3 Installations incorporating a generator
Installations incorporating generator sets are required to comply with Section 551 and the general regulations of BS 7671. Where a generator is used to supply a temporary installation forming part of a TN or TT system, the installation is required to be earthed, preferably by separate earth electrodes. For TN systems all exposed-conductive-parts should be connected by protective conductors to the generator. The neutral conductor and/or star-point of the generator should be connected to the exposed-conductive-parts of the generator and referenced to Earth, as shown in Figure 10.12.

Part VI of the *ESQCR 2002* provides requirements for generation. Regulation 21 sets out requirements for switched alternative sources of energy as follows:

Regulation 21. Where a person operates a source of energy as a switched alternative to a distributor's network, he shall ensure that that source of energy cannot operate in parallel with that network and where the source of energy is part of a low voltage consumer's installation, that installation shall comply with British Standard Requirements.

10.8.4 Final circuits

It is highly desirable that each circuit for socket-outlets, rated up to 32 A, and for each final circuit other than one for emergency lighting, be protected by an RCD with a rated residual operating current not exceeding 30 mA.

10.9 Potentially explosive atmospheres

BS 7671 identifies a number of types of electrical installations for which additional requirements are embodied in other standards. For potentially explosive atmospheres, there are three standards with which compliance is required:

- BS EN 60079: *Electrical apparatus for explosive gas atmospheres.*
- BS EN 50014: *Electrical apparatus for potentially explosive atmospheres.*
- BS EN 50281: *Electrical apparatus for use in the presence of combustible dust.*

With regard to earthing and bonding, the requirements of BS 7671 general apply. However, a TN-C system and a multiple-earthed TN-C-S system are not considered suitable for locations where potentially explosive atmospheres exist.

The Institute of Energy has published a number of guidance publications which will aid electrical installation designers, including:

- *Model code of safe practice Part 15: Area classification code for installations handling flammable fluids*
- *Model code of safe practice Part 2: Design, construction and operation of petroleum distribution installations*
- *Code of safe practice for contractors working on filling stations*
- *Road tanker workshop code*
- *Design, construction, modification, maintenance and decommissioning of filling stations*

Section 14 of the Institute of Energy's publication *Design, construction, modification, maintenance and decommissioning of filling stations* addresses the particular requirements relating to electrical installations in filling stations. Clause 14.3.5 provides recommendations regarding site electricity supplies. Amongst other things, the clause recommends that for a new or refurbished site the installation should be one of the following:

- A TN-S system exclusive to the filling station or exclusive to a larger premises of which the filling station forms part. In either case the premises would have a distribution transformer for its sole use.
- A transformer exclusive to the filling station providing a local TN-S system with its own earth electrode arrangement, independent of the supply network earthing.
- TT system with earthing arrangements exclusive to the filling station.
- Use of an isolation transformer to derive a separate neutral and earth system from a public supply network, as shown in Figure 10.13.

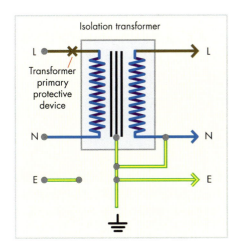

▶ **Figure 10.13** Local TN-S system

The Institute of Energy's publication also addresses:

▶ lightning protection – Clause 14.3.4
▶ protection against static electricity – Clause 14.4.7
▶ cathodic protection – Clause 14.4.8.

The publication also calls for provision of a socket-outlet for the exclusive use for measurement of the external loop impedance, Z_e, and the prospective fault current, I_{pf} (Figure 10.14). The socket-outlet is required to be 'all-insulated' and connected upstream of the main switch for the filling station via an all-insulated means of isolation. Cables for inter-linking these items are required to be insulated and sheathed or insulated and enclosed in a non-metallic conduit, with the length restricted to 3 m. The live conductors should be connected to the incoming side of the installation's main switch.

The protective conductor for the test socket-outlet is required to be segregated from the earthing arrangements of the installation with the earth terminal of the test socket-outlet connected to the earthing conductor's side of the MET. For test purposes the disconnectable link of the MET should be disconnected, having first safely and securely isolated the entire installation. On completion of the measurements, the disconnectable link of the MET should be re-connected before the installation is re-energised.

The isolating device for the test socket-outlet should be equipped with a suitable mechanism to lock it in the OFF position. Additionally, a durable warning notice, as shown in Figure 10.15, is required to be affixed on or in close proximity to the device.

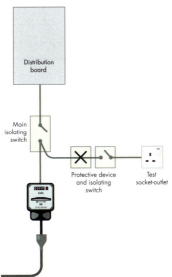

▶ **Figure 10.14** Test socket-outlet

> **Figure 10.15** Warning notice

> This device is not isolated by the main isolating switch and must remain locked or interlocked in the OFF position when not being used for test purposes.

10.10 Medical locations

Chapter 10 of IEE Guidance Note 7 provides advice on electrical installations in medical locations and should be considered in full by those designers engaged in such work. This Guide draws from it the advice given concerning earthing and bonding.

10.10.1 TN-C system

The TN-C system is not permitted in medical locations and medical buildings downstream of the main distribution board. This is prescribed to avoid any possible electromagnetic interference with sensitive medical electrical equipment caused by load currents circulating in PEN conductors and parallel paths.

Regulation 8(4) of *ESQCR 2002* precludes the use of combined neutral and earth conductors in any part of a consumer's installation.

10.10.2 SELV and PELV

Protection by SELV and PELV in medical locations belonging to Group 1 and Group 2 (see below) is limited to 25 V rms AC or 60 V ripple-free DC.

When using SELV and/or PELV, protection by insulation of live parts or by barriers or enclosures is necessary even where the nominal voltage does not exceed 25 V rms AC or 60 V ripple-free DC.

Exposed-conductive-parts of PELV equipment should be connected to the local equipotential bonding conductor connecting all other exposed-conductive-parts and extraneous-conductive-parts.

The groups referred to previously are:

- *Group 0*. Medical locations where no applied parts are intended to be used.
- *Group 1*. Medical locations where applied parts are intended to be used as follows:

    i externally
    ii invasively to any part of the body, except where Group 2 applications are intended.

- *Group 2*. Medical locations where applied parts are intended to be used in applications such as intracardiac procedures, operating theatres and vital treatment where discontinuity (failure) of the supply can cause danger to life.

10.10.3 Protection against indirect contact

Protection by automatic disconnection of supply, by electrical separation or by the use of Class II equipment (or equipment having equivalent insulation) may be used, except as described below:

- *Generally:* In medical locations belonging to Group 1 and Group 2, where TN, TT and IT systems are installed, the conventional touch voltage should not exceed 25 V. Additionally, for TN and IT systems, the resistance of the CPC should not exceed the value given in Table 41C of BS 7671.

- *TN systems:* In medical locations belonging to Group 1 and Group 2, a single-phase final circuit rated up to 32 A is required to incorporate an RCD of Type A or Type B.

The maximum residual operating current should be 30 mA and devices used as follows:

Group 1. 30 mA RCD for all circuits.

Group 2. 30 mA RCD for circuits for:

- the supply to mechanism controlling the manoeuvrability of operating tables only
- X-ray units
- equipment with a rated power exceeding 5 kVA
- non-critical electrical equipment (non life-support).

10.10.4 TT systems
In medical locations belonging to Group 1 and Group 2, the above requirements for TN systems apply and in all cases residual current devices are required to be used.

10.10.5 Medical IT systems
For Group 2 medical locations, the medical IT system (IT electrical system having specific requirements for medical applications) is required to be used for circuits supplying medical electrical equipment and medical systems intended for life support, surgical applications and other electrical equipment located in the patient environment.

Medical IT systems provide both additional protection from electric shock and improved security of supply under single earth fault conditions.

For each group of rooms serving the same function, at least one separate medical IT system is necessary. The medical IT system should be equipped with an insulation monitoring device (IMD), in accordance with IEC 61557-8 (BS EN 61557-8) *Insulation monitoring devices for IT systems.* The IMD is required to incorporate both acoustic and visual alarms situated at a suitable place for permanent monitoring by the medical staff. It should meet the following specific requirements:

- the AC internal resistance should be at least 100 kΩ
- the test voltage should not be greater than 25 V d.c.
- the test current should, even under fault conditions, not be greater than 1 mA peak
- indication should take place at the latest, when the insulation resistance has decreased to 50 kΩ
- a test device should be provided

10.10.6 Transformers for a medical IT system
Transformers for a medical IT system are required to comply with IEC 61558-2-15 (BS EN 61558-2-15). Additional requirements for isolating transformers for the supply of medical locations are:

- The leakage current of the output winding to earth and the leakage current of the enclosure, where measured in a no-load condition with the transformer supplied at rated voltage and rated frequency, should not exceed 0.5 mA.

▶ Single-phase transformers should be used to form the medical IT systems for portable and fixed equipment and the rated output should not be less than 0.5 kVA and should not exceed 10 kVA.

Monitoring for overload conditions and temperatures for the medical IT transformer is required. Figures 10.16 and 10.17 show a typical IT system with insulation monitoring.

Where the supply of three-phase loads via an IT system is required, a separate three-phase transformer should be provided for this purpose with an output line-to-line voltage that does not exceed 250 V.

▶ **Figure 10.16** Typical IT system with insulation monitoring – theatre suite

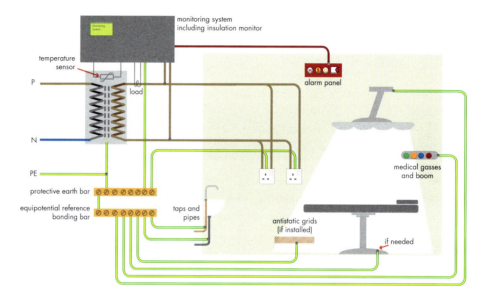

▶ **Figure 10.17** Typical IT system with insulation monitoring – distribution network

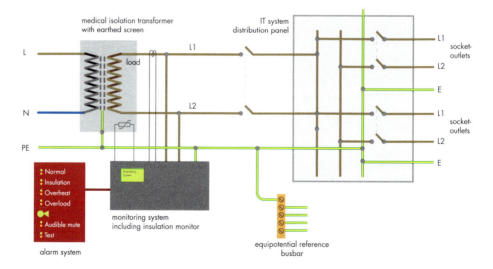

10.10.7 Supplementary equipotential bonding

In each Group 1 and Group 2 medical location, supplementary equipotential bonding conductors should be installed and connected to the equipotential bonding busbar for the purpose of equalising potential differences between the following parts, located in the 'patient environment' (as shown in Figure 10.16):

▶ protective conductors
▶ extraneous-conductive-parts

- screening against electrical interference fields, where installed
- connection to conductive floor grids, where installed
- metal screen of the isolating transformer, if any.

In Group 2 medical locations, the resistance of the conductors, including that of the connections, between the terminals for the protective conductor of socket-outlets and of fixed equipment or any extraneous-conductive-parts and the bonding busbar should not exceed 0.2 Ω.

The equipotential bonding busbar should be located in or near to the medical location. Connections should be so arranged that they are clearly visible and readily disconnectable.

Any wiring system within Group 2 medical locations should be exclusive to the use of equipment and fittings in that location.

10.11 Marinas

Chapter 9 of IEE Guidance Note 7 provides advice on electrical installations in marinas and should be considered in full by those designers engaged in such work. This Guide draws from it the advice given concerning earthing and bonding.

The guidance given in Chapter 9 of IEE Guidance Note 7 applies to the electrical installations of marinas providing facilities for the supply of electricity to leisure craft, in order to provide for standardisation of power facilities. The requirements do not apply to the electrical installations in offices, workshops, toilets, leisure accommodation etc. which form part of the marina complex, where the general requirements of BS 7671 apply.

10.11.1 The risks

The environment of a marina or yachting harbour is harsh for electrical equipment. The water, salt and movement of structures accelerate deterioration of the installation. The presence of salt water, dissimilar metals and a potential for leakage currents increases the level of corrosion. There are also increased risks of electric shock associated with a wet environment, by reduction in body resistance and contact with earth potential.

The risks specifically associated with craft supplied from marinas include:

- Open circuit faults of the PEN conductor of PME supplies raising the potential to true earth of all metalwork (including that of the craft, where connected) to dangerous levels.
- Inability to establish an equipotential zone external to the craft.
- Possible loss of earthing due to long supply cable runs, connecting devices exposed to weather and flexible cord connections liable to mechanical damage.

10.11.2 Minimising the risks

Particular requirements to reduce the above risks include:

- Prohibition of the connection of exposed-conductive-parts and extraneous-conductive-parts of the craft to a PME terminal (where such an earthing facility is made available by the electricity distributor).
- Additional protection by 30 mA RCDs in both the craft and the marina installation.

10.11.3 Protection against direct contact

An RCD that complies with BS 4293, BS 7071, BS 7288, BS EN 61008-1 or BS EN 61009-1 may be used to reduce the risk of electric shock caused by direct contact. However, an RCD should not be used as the sole means of protection against direct contact.

10.11.4 Protection against indirect contact

Where protection by automatic disconnection of supply is selected:

- for TN systems, only a TN-S arrangement may be used
- for TT systems, an RCD that disconnects phase and neutral conductors complying with BS EN 61008-1, BS 4293 or BS EN 61009-1, and having the characteristics specified in Regulation 412-06 should be used, including where protection is provided by an onshore isolating transformer.

Only permanent onshore buildings may use the electricity distributor's PME earthing terminal. For the boat mooring area of the marina this is not permissible, and entirely separate earthing arrangements are required to be provided. This is generally achieved by the use of a suitably rated RCD complying with BS EN 61008 with driven earth rods or mats providing a TT system for that part of the installation.

Marina installations are often of sufficient size to warrant the provision of an 11 kV/415 V transformer substation. In these, and sometimes in other, circumstances the electricity distributor may be willing to provide a TN-S supply, which is much more suitable for such installations. Where the transformer belongs to the marina, a TN-S system should be installed.

10.11.5 Isolating transformers

Supplies to craft may be provided from any supply system through isolating transformers. This method has the advantage of reducing electrolytic corrosion and can be used with TN-S and TN-C-S (PME) supplies.

The isolating transformer is required to comply with BS EN 60742: *Isolating transformers and safety isolating transformers*, or the BS EN 61558 series of standards. See Figures 10.18 to 10.20 for typical wiring arrangements with onshore mounted isolating transformers.

Connection of the protective conductor of the shore supply should not be made to the bonding of the leisure craft. However, the following items should be effectively and reliably connected to a bonding conductor – which, in turn, should be connected to one of the secondary winding terminals of the isolating transformer:

- Metal parts of the leisure craft which are in electrical contact with water. Where the type of construction does not ensure continuity, then more than one connection point may be required.
- The protective contact of each socket-outlet.
- The exposed-conductive-parts of electrical equipment.

Only one craft (socket-outlet) should be connected to each secondary winding of an isolating transformer.

The isolating transformer isolates the craft installation from the shore, allowing supplies to be taken from multiple-earth networks, and provides some protection against electrolytic corrosion. It does not provide protection against direct contact or indirect contact. The craft earth is connected to one pole of the secondary isolating transformer.

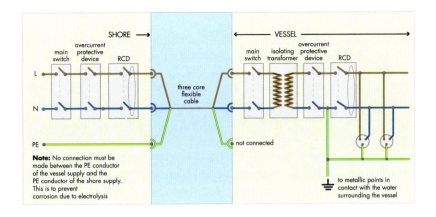

▶ **Figure 10.18**
Connection to mains supply with an RCD

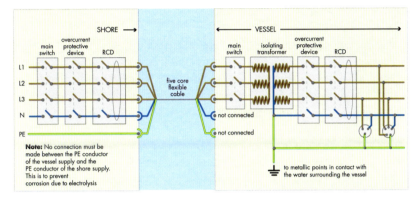

▶ **Figure 10.19**
Connection to mains supply with a three-phase socket-outlet

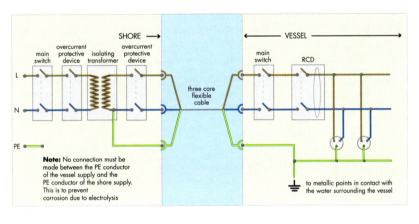

▶ **Figure 10.20**
Onshore mounted isolating transformer (hull and metal parts bonded)

10.12 Cable tray and cable basket

A cable tray or a cable basket where used as a support and cable management system has to be considered in the context of earthing and bonding. In other words, are such systems, where consisting of metal and plastic-coated metal, exposed-conductive-parts or extraneous-conductive-parts and consequently do they require earthing or bonding?

Addressing, first, the question of earthing and whether the cable tray or basket should be earthed, electrical equipment such as cables mounted on a metallic support system will normally be equivalent to either a Class I construction (for example copper sheathed, mineral insulated cables without an overall PVC covering) or a Class II equivalent construction (for example PVC insulated and sheathed cable).

Exposed-conductive-parts of cables, such as the copper sheath of a mineral insulated cable, are required to be connected to the MET of the installation by a CPC designed to conduct earth fault currents. The cable tray or basket which the mineral insulated cable is attached to, or may be in contact with, is not itself an exposed-conductive-part and therefore it does not require earthing. To do so would only serve to distribute further any touch voltage resulting from an earth fault on an item of equipment to which the cable is connected.

A cable complying with the appropriate standard having a non-metallic sheath or a non-metallic enclosure is deemed to provide satisfactory protection against both direct and indirect contact, as does an item of Class II equipment (Regulation 471-09-04 refers). Class II equipment is constructed such that any insulation fault in the cable cannot result in a fault current flowing into any conductive parts with which the equipment may be in contact. Hence, the metal cable tray or basket need not be earthed (Figure 10.21).

Generally, the conductive parts of a metal cable tray or basket system need not be purposely earthed. Some conductive parts of a metal cable support system may be earthed, however, by virtue of fortuitous contact with exposed-conductive-parts.

Should the cable tray or basket be equipotentially bonded? Unless the metal cable support system introduces a potential that does not already exist in the location in which the system is installed, it will not meet the definition of an extraneous-conductive-part. In normal circumstances therefore there is no need to arrange for the conductive parts of the support system to be connected to either a main bonding conductor or any supplementary bonding conductor.

However, should the cable tray be installed in such a manner that it is likely to introduce a potential from outside the location, thereby meeting the definition of an extraneous-conductive-part, then main equipotential bonding will be required (Regulation 413-02-02 refers). For example, consider a run of cable tray carrying services into a particular building. The cable tray may be in contact with the earth potential outside a building and upon entering into the building would be likely to introduce the earth potential into that building. In such a case the tray would warrant the definition of an extraneous-conductive-part and consequently main equipotential bonding would be required.

Where, for example, the installation designer has selected a cable tray for use as a protective conductor, which is permitted under Regulation 543-02-02(vi) where it is described as an electrically continuous support system for conductors, the cable tray would be required to meet the requirements for a protective conductor given in Regulation 543-02-04 and would need to be connected with earth.

▶ **Figure 10.21** Cable tray

10.13 Inspection and testing of protective conductors

Whilst live conductors are required for functional purposes, protective conductors are essential for safety and it is always necessary to inspect and test them to ensure that they will perform as intended. The following check list may assist in undertaking the necessary inspections:

- A satisfactory means of earthing exists.
- The MET has been provided and is readily accessible (Regulation 542-04-01).
- Provision for disconnecting the earthing conductor (Regulation 542-04-02).
- Exposed-conductive-parts have been earthed (Regulations 413-02-06, 413-02-18 and 413-02-23).
- Circuits suitably identified (neutral and protective conductors in same sequence as phase conductors) (Regulations 514-01-02 and 514-08-01).
- Correct cable glands and gland-plates used (BS 6121).
- Cables used comply with British or Harmonised Standards (Appendix 4 of the Regulations, Regulation 521-01-01).
- Earth tail pots installed where required on mineral insulated cables (Regulation 133-01-04).
- Non-conductive finishes on enclosures removed to ensure good electrical connection and where necessary made good after connecting (Regulation 526-01-01).
- Consideration paid to electromagnetic effects and electromechanical stresses (Chapter 52).
- Protective conductors securely fixed (Regulation 133-01-01).
- Protective conductors suitably labelled (Regulation 514-01).
- Non-conductive finishes on switchgear removed at protective conductor connections and where necessary made good after connecting (Regulation 526-01-01).
- Suitable cable glands and gland plates used (Regulation Group 526-01).
- External influences likely to be encountered taken account of, i.e. suitable for the foreseen environment (Section 522).
- Correct IP rating applied (BS EN 60529).
- Protective conductors correctly terminated and identified (Sections 514 and 526).
- All connections mechanically secure and electrically sound (Section 526).
- Protective conductors correctly identified (Regulation 514-06).
- Bare protective conductors sleeved green-and-yellow (Regulation 543-03-02).
- Terminals tight and containing all strands of the conductors (Section 526).
- Earthing of all exposed-conductive-parts, e.g. metal switchplate (Chapter 54).
- CPC connected directly to the earthing terminal of the socket-outlet, on a sheathed wiring installation (Regulation 543-02-07).
- Earthing tail from the earthed metal box, on a conduit installation to the earthing terminal of the socket-outlet (Regulation 543-02-07).
- Complies with the requirements for the special locations where applicable.
- Metal wiring systems earthed.
- High-integrity protective conductor(s) where applicable.
- Metal sheaths and armouring earthed (Regulation 543-02).
- Glands correctly selected and fitted with shrouds and supplementary earth tags as necessary (Regulation 526-01).
- Earthed concentric wiring including CNE cables to be used only as permitted.
- Protected where exposed to mechanical damage (Regulations 522-06 and 522-08).
- Prohibited core colours not used.
- Joints in metal conduit, duct or trunking compliant with BS 7671 (Regulation 543-03).
- Flexible conduit supplemented by a protective conductor (Regulation 543-02-01).
- Minimum CSA of protective conductors (543-01).

- Copper conductors, other than strip, of 6 mm² or less protected by insulation (Regulation Group 543-03).
- CPC at termination of sheathed cables insulated with sleeving (Regulation 543-03-02).
- Bare CPC protected against mechanical damage and corrosion (Regulations 542-03 and 543-03-01).
- Insulation, sleeving and terminations identified by colour combination green and yellow (Regulation 514-03-01).
- Main and supplementary bonding conductors of sufficient CSA (Section 547).
- Separate CPCs not less than 4 mm² where not protected against mechanical damage (Regulation 543-01-01).

The above check list is not exhaustive and the order of items does not signify their relative importance – they are all important where relevant to the particular installation.

Values for *k* for various forms of protective conductor

Table 43A of BS 7671
Values of *k* for common materials, for calculation of the effects of fault current

Conductor material	Insulation material	Assumed initial temperature (°C)	Limiting final temperature (°C)	*k*
Copper	70 °C thermoplastic (general purpose PVC)	70	160/140*	115/103*
	90 °C thermoplastic (PVC)	90	160/140*	100/86*
	60 °C thermosetting (rubber)	60	200	141
	85°C thermosetting (rubber)	85	220	134
	90 °C thermosetting	90	250	143
	Impregnated paper	80	160	108
Copper	Mineral – plastic covered or exposed to touch	70 (sheath)	160	115
	Mineral – bare and neither exposed to touch nor in contact with combustible materials	105 (sheath)	250	135
Aluminium	70 °C thermoplastic (general purpose pvc)	70	160/140*	76/68*
	90 °C thermoplastic (pvc)	90	160/140*	66/57*
	60 °C thermosetting (rubber)	60	200	93
	85 °C thermosetting (rubber)	85	220	89
	90 °C thermosetting	90	250	94
	Impregnated paper	80	160	71

Notes:
These data are applicable only for disconnection times up to 5 s. For longer times the cable manufacturer shall be consulted.
* Where two values of limiting final temperature and of *k* are given, the lower value relates to cables having conductors of greater than 300 mm^2 CSA.

Table 54B of BS 7671
Values of k for insulated protective conductor not incorporated in a cable and not bunched with cables, or for separate bare protective conductor in contact with cable covering but not bunched with cables where the assumed initial temperature is 30 °C

	Insulation of protective conductor or cable covering			
	70 °C thermoplastic (general purpose PVC)	90 °C thermoplastic (PVC)	85 °C thermosetting (rubber)	90 °C thermosetting
Conductor material: Copper	143/133*	143/133*	166	176
Conductor material: Aluminium	95/88*	95/88*	110	116
Conductor material: Steel	52	52	60	64
Final temperature (°C)	160/140*	160/140*	220	250

*Above 300 mm^2

Table 54C of BS 7671
Values of k for protective conductor incorporated in a cable or bunched with cables, where the assumed initial temperature is 70 °C or greater

	Insulation material			
	70 °C thermoplastic (general purpose PVC)	90 °C thermoplastic (PVC)	85 °C thermosetting (rubber)	90 °C thermosetting
Conductor material: Copper	115/103*	100/86*	134	143
Conductor material: Aluminium	76/68*	66/57*	89	94
Assumed initial temperature (°C)	70	90	85	90
Final temperature (°C)	160/140*	160/140*	220	250

*Above 300 mm^2

Table 54D of BS 7671
Values of k for protective conductor as a sheath or armour of a cable

	Insulation material			
	70 °C thermoplastic (general purpose PVC)	90 °C thermoplastic (PVC)	85 °C thermosetting (rubber)	90 °C thermosetting
Conductor material: Aluminium	93	85	93	85
Conductor material: Steel	51	46	51	46
Conductor material: Lead	26	23	26	23
Assumed initial temperature (°C)	60	80	75	80
Final temperature (°C)	200	200	220	200

Table 54E of BS 7671
Values of k for steel conduit, ducting and trunking as the protective conductor

	Insulation material			
	70 °C thermoplastic (general purpose PVC)	90 °C thermoplastic (PVC)	85 °C thermosetting (rubber)	90 °C thermosetting
Conductor material: Steel conduit, ducting and trunking	47	44	54	58
Assumed initial temperature (°C)	50	60	58	60
Final temperature (°C)	160	160	220	250

Data for armoured cables   B

CSA of phase conductor (S) (mm²)	Table 54G formula (BS 7671)	Minimum required CSA of armouring (mm²)	Actual CSA of armouring from BS 5467 and BS 6724		
			2-core (mm²)	3-core (mm²)	4-core (mm²)
1.5	$\frac{k_1}{k_2} \times S$	4.66	15	16	17
2.5		7.77	17	19	20
4		12.43	19	20	22
6		18.65	22	23	36
10		31.09	26	39	42
16		49.74	42	45	50
25	$\frac{k_1}{k_2} \times 16$	49.74	42	62	70
35		49.74	60	68	78
50	$\frac{k_1}{k_2} \times \frac{S}{2}$	77.72	68	78	90
70		108.8	80	90	131
95		147.7	113	128	147
120		186.5	125	141	206
150		233.2	138	201	230
185		287.6	191	220	255
240		373.1	215	250	289
300		466.3	235	269	319
400		621.7	265	304	452[4]

▶ **Table B1** Multi-core armoured cables having thermosetting insulation (copper conductors and steel wire armouring). BS 5467 or BS 6724 (90 °C)

Notes:
1 Data in red indicates that the CSA is insufficient to meet BS 7671 Table 54G requirements.
2 $k_1 = 143$. This is the value of k for the phase conductor, selected from Table 43A of BS 7671 (copper conductor with 90 °C thermosetting insulation).
3 $k_2 = 46$. This is the value of k for the protective conductor, selected from Table 54D of BS 7671 (steel armouring with 90 °C thermosetting insulation).
4 Not applicable to cables to BS 6724.

▶ **Table B2** Multi-core armoured cables having thermosetting insulation (copper conductors and steel wire armouring). BS 5467 or BS 6724

NB: DATA TO BE USED ONLY WHERE THE CONDUCTOR OPERATING TEMPERATURE WILL NOT EXCEED 70 °C

CSA of phase conductor (S) (mm²)	Table 54G formula (BS 7671)	Minimum required CSA of armouring (mm²)	Actual CSA of armouring from BS 5467 and BS 6724		
			2-core (mm²)	3-core (mm²)	4-core (mm²)
1.5	$\frac{k_1}{k_2} \times S$	3.38	15	16	17
2.5		5.64	17	19	20
4		9.02	19	20	22
6		13.53	22	23	36
10		22.55	26	39	42
16		36.08	42	45	50
25	$\frac{k_1}{k_2} \times 16$	36.08	42	62	70
35		36.08	60	68	78
50		56.37	68	78	90
70		78.92	80	90	131
95		107.11	113	128	147
120	$\frac{k_1}{k_2} \times \frac{S}{2}$	135.29	125	141	206
150		169.12	138	201	230
185		208.58	191	220	255
240		270.59	215	250	289
300		338.24	235	269	319
400		450.98	265	304	452[4]

Notes:
1 Data in red indicates that the CSA is insufficient to meet BS 7671 Table 54G requirements.
2 $k_1 = 115$. This is the value of k for the phase conductor, selected from Table 43A of BS 7671 (copper conductor with assumed initial temperature of 70 °C and assumed limiting final temperature of 160 °C).
3 $k_2 = 51$. This is the value of k for the protective conductor, selected from Table 54D of BS 7671 (steel armouring with assumed initial temperature of 60 °C and assumed limiting final temperature of 200 °C).
4 Not applicable to cables to BS 6724.

CSA of phase conductor (S) (mm²)	Table 54G formula (BS 7671)	Minimum required CSA of armouring (mm²)	Actual CSA of armouring from BS 5467 and BS 6724		
			2-core (mm²)	3-core (mm²)	4-core (mm²)
16	$\frac{k_1}{k_2} \times S$	32.7	39	41	46
25		51.09	38	58	66
35	$\frac{k_1}{k_2} \times 16$	32.7	54	64	72
50		32.70	60	72	82
70		71.52	*70*	84	122
95		97.07	100	119	135
120		122.61	-	131	191
150	$\frac{k_1}{k_2} \times \frac{S}{2}$	153.26	-	181	211
185		189.26	-	206	235
240		245.22	-	*230*	265
300		306.52	-	*250*	*289*

▶ **Table B3** Multi-core armoured cables having thermosetting insulation (solid aluminium conductors and steel wire armouring). BS 5467 or BS 6724 (90 °C)

Notes:
1 Data in red indicates that the CSA is insufficient to meet BS 7671 Table 54G requirements.
2 $k_1 = 94$. This is the value of k for the phase conductor, selected from Table 43A of BS 7671 (aluminium conductor with 90 °C thermosetting insulation).
3 $k_2 = 46$. This is the value of k for the protective conductor, selected from Table 54D of BS 7671 (steel armouring with 90 °C thermosetting insulation).

NB: DATA TO BE USED ONLY WHERE THE CONDUCTOR OPERATING TEMPERATURE WILL NOT EXCEED 70 °C

CSA of phase conductor (S) (mm²)	Table 54G formula (BS 7671)	Minimum required CSA of armouring (mm²)	Actual CSA of armouring from BS 5467 and BS 6724		
			2-core (mm²)	3-core (mm²)	4-core (mm²)
16	$\frac{k_1}{k_2} \times S$	23.84	39	41	46
25		23.84	38	58	66
35	$\frac{k_1}{k_2} \times 16$	23.84	54	64	72
50		37.25	60	72	82
70		52.16	70	84	122
95		70.78	100	119	135
120	$\frac{k_1}{k_2} \times \frac{S}{2}$	89.41	-	131	191
150		111.76	-	181	211
185		137.84	-	206	235
240		178.82	-	230	265
300		223.53	-	250	289

▶ **Table B4** Multi-core armoured cables having thermosetting insulation (solid aluminium conductors and steel wire armouring). BS 5467 or BS 6724

Notes:
1 $k_1 = 76$. This is the value of k for the phase conductor, selected from Table 43A of BS 7671 aluminium conductor with assumed initial temperature of 70 °C and assumed limiting final temperature of 160 °C).
2 $k_2 = 51$. This is the value of k for the protective conductor, selected from Table 54D of BS 7671 (steel armouring with assumed initial temperature of 60 °C and assumed limiting final temperature of 200 °C).

Data for armoured cables

Index

access doors, street furniture 140
accessories
　terminations in 113–14, 119–20
　see also socket-outlets
additions to installation 44–5
advertising signs 140
agricultural premises
　automatic disconnection 65
　earth electrode resistance 17
　supplementary equipotential bonding 87, 94–7
alterations to installation 44–5
alternative supplies, automatic disconnection 79
aluminium conductors
　armouring, termination 108
　circuit protective conductors 104–6
　k values 155–6
　main equipotential bonding conductors 42
　resistivity 126
ambient temperature 68
ambulances 137
armouring, cable
　circuit protective conductors 106–8, 121
　earthing and bonding clamps 58
　earthing conductor 21
　k values 156
　main equipotential bonding conductors 40
arm's reach 55, 139
automatic disconnection 4, 63–81
　and alternative supplies 79
　fixed equipment outdoors 75
　functional extra-low voltage systems 81
　IT systems 73–5
　portable equipment used outdoors 75
　protective extra-low voltage systems 80–1
　RCDs in series 75–6
　reduced low voltage circuits 77–9
　separated extra-low voltage systems 80
　TN systems 63–73
　TT systems 73
　see also earthed equipotential bonding and
　　automatic disconnection of supply (EEBADS)
automatic disconnection not achievable 99–100

back boxes 113–14
bathrooms 88–91
blocks of flats 52–3
boats 137
　marinas 149–51
　protective multiple earthing 130
body impedance 56, 117
boiler interiors 97
bonding clamps 58–61
bonding conductors: see main equipotential bonding
　　conductors; supplementary bonding conductors
broadcasting vehicles 137, 138
BS 1361 66, 67, 68
BS 1362 66, 67, 68
BS 3036 66, 67, 68
BS 3636 77
BS 4293 75, 143, 150
BS 4444 121, 122, 123
BS 4568 110
BS 4678 112
BS 4884 2
BS 4940 2
BS 5467 157–9
BS 6004 86, 126
BS 60898 78
BS 6121 153
BS 6651 46
BS 6724 157–9
BS 7071 75, 150
BS 7211 86, 126
BS 7288 75, 150
BS 729 7
BS 7430 1, 6–7, 9, 11–12, 14, 28, 61
BS 7671 1, 27, 49, xi
BS 8290 141
BS 88 37, 66, 67, 68, 78, 99, 103, 109
BS 951 58–61
BS EN 10025 7
BS EN 50014 144
BS EN 50281 144
BS EN 60079 144
BS EN 60309 118, 120, 121
BS EN 60439 113
BS EN 60529 153
BS EN 60742 77, 150
BS EN 60898 67, 71, 72, 78
BS EN 60950 118
BS EN 61008 73, 75, 143, 150
BS EN 61009 67, 71, 73, 75, 78, 143, 150
BS EN 61557 147
BS EN 61558 147

BS PD 6519 56
buried earthing conductor, cross-sectional area 23–4
buried steel grids, connections to 61–2
busbar systems, circuit protective conductors 120
bus shelters 140

cabins, transportable 137
cable armouring: *see* armouring, cable
cable baskets 151–2
cable sheathing: *see* sheathing, cable
cable trays 151–2
caravan parks 130–1
caravans 137
car park ticket dispensers 140
catenary wires, cable supports 142
cathodic protection, petrol filling stations 145
chicken-houses 94
circuit-breakers, maximum earth fault loop
    impedance 67, 71, 72, 78
circuit protective conductors 99–100
    armouring of cables 106–8
    cross-sectional areas 102–6
    earth monitoring 122–4
    functional earthing 115
    installations serving more than one building 50–2
    maximum impedance 68–72
    metal enclosures 112
    proving continuity 124–8
    significant conductor currents 115–22
    steel conduit 108–11
    steel trunking and ducting 111–12
    as supplementary equipotential bonding
        conductor 85
    terminations in accessories 113–14
Class II equipment, supplementary equipotential
    bonding 89, 90
clean earths 129
Code of practice for earthing: see BS 7430
*Code of safe practice for contractors working on filling
    stations* 144
colour identification
    bonding conductors 42–3
    clean earths 129
    earthing and bonding clamps 60
    earthing conductor 24
    functional earthing 115
combined earthing arrangement
    TN-C-S system 30–1
    TN-C system 28, 115
combined heat and power 133
computing equipment, clean earths 129
conductance
    equivalent 41–2
    supplementary bonding conductors 86
conductivity 55
    see also resistivity
connections
    CPC current exceeding 10mA 120–1

earthing conductor 25
    extraneous-conductive-parts 58–62
    lightning protection systems 46–7
    main equipotential bonding 39–40
    metallic grids 93–4, 96–7
    see also terminations
construction sites
    automatic disconnection 65
    cabins 137
    temporary supplies 141
continuity
    proving 124–8
    steel conduit 111
    steel trunking 112
controlgear, metal enclosure or frame as CPC
    112–13
copper conductors
    circuit protective conductors 103–4
    galvanic corrosion 14
    k values 155–6
    main equipotential bonding conductors 41
    resistance 66, 87
    resistivity 126
    rod electrodes 6–7
    supplementary bonding conductors 87, 88
    supports 88
    tape and wire electrodes 8
corrosion protection
    circuit protective conductors 101
    earth electrodes 14
    earthing conductor 24–5
cow houses 95–6
cross-sectional area
    circuit protective conductors 102–6
        armoured cable 106, 107, 157–9
        busbar systems 120
        significant conductor currents 117, 120–1
        steel conduit 109–10
        steel trunking 111–12
    earthing conductor 22–4
    main equipotential bonding conductors 41–2
    rod electrodes 6–7
    supplementary bonding conductors 86
    tape and wire electrodes 8

dairies 95–6
*Design, construction, modification, maintenance and
    decommissioning of filling stations* 144
direct contact 64
    marinas 150
    mobile and transportable units 139
    portable equipment for use outdoors 75
    street furniture 140
disconnection of the earthing conductor
    21–2, 25, 145
disconnection times 63
    IT systems 74
    TN systems 64–5, 65–8, 72

TT systems 73
distribution boards
   highways 139
   labelling 118
distribution circuits
   automatic disconnection unachievable 99–100
   exhibitions, shows and stands 143
   extraneous-conductive-parts common to a number of buildings 47–9
   highways 140
   installations serving more than one building 49–52
   mixed disconnection times 69–72
   multiple-occupancy premises 52–3
distributors' facilities 5–6, 19

earth bar 48–9
earthed equipotential bonding and automatic disconnection of supply (EEBADS) 4, 36–8, 63, 83
earth electrodes 6
   earth plates 8–9
   installation 14–15
   loading capacity 15
   location 10
   metal covering of cables 10
   resistance
      calculation 10–14
      external earth fault loop impedance 18–19
      measurement 15–18
   responsibility for providing 19
   structural metalwork electrodes 9–10
   testing 15–18
earth fault loop
   IT systems 34
   TN-C-S systems 31
   TN-C systems 29
   TN-S systems 30
   TT systems 33
earth fault loop impedance
   contribution of cable armouring as CPC 107
   IT systems 74
   reduced low voltage systems 78–9
   test instrument 17–18
   TN systems 65–9
   TT systems 73
earth-free equipotential bonding 36, 45–6, 143
earthing arrangements, system types 27–34
earthing clamps 58–61
earthing conductor 21–5
   colour identification 24
   connection to means of earthing 21–2
   connection to means of testing 25
   cross-sectional area 22–4
   disconnection 21–2, 25
   galvanic corrosion 14
   impedance contribution 24
   protection against external influences 24–5

terminations
   earth plates 9
   rod electrodes 7–8
   structural metalwork electrodes 9–10
earthing tails 113–14
earthing terminal: *see* main earthing terminal
earth marshalling bar, main equipotential bonding 48–9
earth monitoring 121, 122–4
'earth noise' 129
earth plates: *see* plate electrodes
earth proving systems 122–4
EEBADS: *see* earthed equipotential bonding and automatic disconnection of supply
electrical separation
   medical locations 146
   mobile and transportable units 138
   portable equipment for use outdoors 75
   restrictive conductive locations 98
   see also isolating transformers
Electricity at Work Regulations 1989 xi
   mobile and transportable units 138
   protective conductor connections 107
Electricity Safety, Quality and Continuity Regulations 2002 (ESQCR)
   caravan parks 130
   exhibitions, shows and stands 143
   IT systems 34
   medical locations 146
   mobile and transportable units 138
   responsibility for providing earthing 19
   small-scale embedded generators 133, 134
   source earthing 28
   TN-C systems 29, 115
Electricity Supply Regulation 1988, main equipotential bonding conductors 45
electric shock 56, 94, 116–17
   see also locations of increased risk
electric vehicles 137
electrodes, earthing: *see* earth electrodes
EMB: *see* earth marshalling bar
embedded generation 133–6
EN 60309 118
enclosures: *see* metal enclosures
Engineering Guidance G83 133–4, 135
equivalent conductance 41–2
ESQCR: *see* Electricity Safety, Quality and Continuity Regulations 2002 (ESQCR)
exhibitions 143–4
explosive atmospheres 144–6
exposed-conductive-parts 4, 27, 64
   touch voltages 35–8
extensions to installation 44–5
exterior gas meters 131–2
external earth fault loop impedance 18–19, 21, 145
external influences, protection of earthing conductor 24–5
extra-low voltage systems 80–1

extraneous-conductive-parts 55–62
   bath and shower rooms 89
   common to a number of buildings 47–9
   connections 58–62
   definition 55–7
   examples 57–8
   installations serving more than one building 49–52
   main equipotential bonding 38–40
   supplementary equipotential bonding 83–5
   swimming pools 92
   touch voltages 35–8

Faraday cage 45
farms: *see* agricultural premises
fault current measurement, petrol filling stations 145
feed-processing locations 94
FELV (functional extra-low voltage systems) 81
fertilizer storage 94
filters, protective conductor currents 116
fire-engines 137
fixed equipment outdoors 75
footpath lighting 139
fuel cells 133
functional earthing 115
   clean earths 129
   restrictive conductive locations 98
functional extra-low voltage systems (FELV) 81
fuses
   maximum earth fault loop impedance 66, 67, 78
   time/current characteristics 109

galvanic corrosion: *see* corrosion protection
gas meters 131–2
gas pipes
   non-use as earth electrodes 7
   non-use as equipotential bonding
      conductors 40, 85
   see also pipework
Gas Safety (Installation and Use)
   Regulations 1994 132
generators
   exhibitions, shows and stands 143–4
   small-scale embedded 133–6
   transportable units 137
gland plates 108
grain silos 97
grids: *see* metallic grids
Guidelines for the design, installation, testing and
   maintenance of main earthing systems in
   substations 28

hand-held equipment 64, 68, 72, 77
hay storage 94
Health and Safety (Safety Signs and Signals)
   Regulations 1996 135
high protective conductor current 115–22
highway power supplies 139–41
horticultural premises
   automatic disconnection 65
   earth electrode resistance 17
   supplementary equipotential bonding 94–7
hydro generation 133

identification: *see* colour identification
IEC 364 136
IEC 479 116–17
IEC 60479 56
IEC 61557 147
IEC 61558 147
IEC Standard 60364-7-11 143
IEE Guidance Note 7
   exhibitions, shows and stands 143
   marinas 149
   medical locations 146–9
   special locations 136, 139
impedance contribution of earthing conductor 24
increased risk: *see* locations of increased risk
indirect contact 4, 36, 56–7
   marinas 150
   medical locations 146–7
   mobile and transportable units 139
industrial kitchens, circuit protective conductors 101
inspection, protective conductors 153–4
installation earth electrode: *see* earth electrodes
Institute of Energy, guidance publications 144, 145
insulation monitoring
   medical IT systems 147, 148
   TT systems 74
isolating transformers
   loop impedance 79
   marinas 150–1
   medical locations 147
   mobile and transportable units 138
   petrol filling stations 144–5
   reduced low voltage systems 77
   significant CPC currents 121, 122
IT equipment
   busbar systems 120
   clean earths 129
   protective conductor currents 116
   significant CPC currents 118
IT systems 27, 34
   automatic disconnection 73–5
   CPC current exceeding 3.5mA 122
   means of earthing 5
   medical locations 147–8
   source earthing 3, 4

k values 22–3, 155–6

labels
   earth-free equipotential bonding 46
   earthing and bonding clamps 58, 60
   earthing conductor 25
   functional earthing 115
   significant CPC currents 118

laundries, circuit protective conductors 101
lead conductors
    circuit protective conductors 104–6
    galvanic corrosion 14
    k values 156
lightning protection systems
    main equipotential bonding 46–7
    petrol filling stations 145
limitations on resistance
    earth electrode 10–14, 17–18
    supplementary bonding conductors 87
        agricultural and horticultural premises 97
        bath and shower rooms 90
        swimming pools 92
livestock 94–6
locations of increased risk
    agricultural premises: *see* agricultural premises
    automatic disconnection 65
    bath and shower rooms 88–91
    non-conducting locations 46
    other locations 99
    restrictive conductive locations 97–8
    supplementary equipotential bonding 83
    swimming pools 91–4
lofts 94
low-noise earth: *see* clean earths
luminaires, protective conductor currents 116

machinery, mobile 137
main earthing terminal (MET) 5
main equipotential bonding 35–53
    earth-free equipotential bonding 45–6
    extraneous-conductive-parts common to a number of buildings 47–9
    installations serving more than one building 49–52
    lightning protection systems 46–7
    multi-occupancy premises 52–3
    purpose 35–8
    typical layout 57
main equipotential bonding conductors 38–45
    alterations and extensions 44–5
    cross-sectional area 41–2
    exterior gas meters 132
    identification 42–3
    supports 43–4
    types 40
marinas 137, 149–51
maximum earth fault loop impedance 65–72
    reduced low voltage circuits 78
means of earthing 5–19
    distributors' facilities 5–6
    responsibility for providing 19
    see also system types
measurements
    continuity 124–8
    earth fault loop impedance 18–19, 65, 145
    resistance of earth electrode 15–18

mechanical protection, earthing conductor 24–5
medical locations 146–9
metal-cladding 58
metal conduits
    circuit protective conductors 101, 108–11
    main equipotential bonding conductors 40
metal ducting
    circuit protective conductors 101, 111–12
    k values 156
metal enclosures, circuit protective conductors 112–13
metal frames, circuit protective conductors 112–13
metallic grids
    connections 93–4, 96–7
    stock houses 96
    swimming pools 93–4
    underfloor bathroom heating 91
metal pipework
    common to a number of buildings 47–9
    extraneous-conductive-parts 56
    restrictive conductive locations 97
    supplementary equipotential bonding 89, 90
    see also pipework
metal trunking
    circuit protective conductors 101, 111–12
    earthing conductor 25
    k values 156
    main equipotential bonding conductors 40
metal window frames 57–8
MET (main earthing terminal) 5
military units 137
milking parlours 95
mixed disconnection times 69–72
mobile units 136–9
mobile workshops 137
Model code of safe practice 144
multi-occupancy premises 52–3

non-conducting locations 45–6, 143
non-metallic enclosures 113
non-metallic pipework
    extraneous-conductive-parts 55
    supplementary equipotential bonding 90

office blocks 52–3
office equipment, significant CPC currents 118
oil pipes
    non-use as earth electrodes 7
    non-use as equipotential bonding conductors 40, 85
    see also pipework
operational manuals 1–2

parallel operation
    photovoltaic generation 136
    small-scale embedded generators 133, 134
    static inverters 99
PELV: *see* protective extra-low voltage systems (PELV)
petrol filling stations 144–6

photovoltaic generation 133, 136
physiological effects 56, 117
piggeries 94
pipework
   common to a number of buildings 47–9
   connections to 58–61
   equipotential bonding conductors 40, 85
   extraneous-conductive-parts 55–6
   restrictive conductive locations 97
   supplementary equipotential bonding 89, 90
placing out of reach 55, 139
plate electrodes 8–9, 10
PME: *see* protective multiple earthing
PNB (protective neutral bonding) 32
polarity testing 127
portable equipment
   disconnection times 64, 68
   outdoors 75
   see also hand-held equipment
potentially explosive atmospheres 144–6
protective conductors: *see* circuit protective conductors
protective earthing 3–4
protective extra-low voltage systems (PELV)
   automatic disconnection 80–1
   medical locations 146
protective multiple earthing (PME) 6, 31–2
   caravan parks 130–1
   main equipotential bonding conductors 40, 42, 45
   mobile and transportable units 138
protective neutral bonding (PNB) 32
proving continuity 124–8
purpose of bonding 35–8
purpose of earthing 4

radial final circuits
   circuit protective conductors 119
   proving continuity 124–6
RCBOs, maximum earth fault loop impedance 67, 71, 78
RCDs 122
   automatic disconnection 73
   earth electrodes
      limit on resistance 13–14, 18
      loading capacity 15
      testing 17–18
   portable equipment used outdoors 75
   reduced low voltage circuits 79
   in series 76–7
reduced low voltage systems 77–9
residual current devices: *see* RCDs
resistance of conductors 65–6
   circuit protective conductors 124–6, 127
   supplementary bonding conductors 87
      agricultural and horticultural premises 97
      bath and shower rooms 90
      swimming pools 92
resistance of earth electrode
   calculation 10–14
   measurement 15–18
   plate electrodes 10, 13
   rod electrodes 10, 12–13
   structural metalwork electrodes 9
   tape electrodes 13
resistivity, copper and aluminium conductors 126
responsibility for earthing 5–6, 19, 28
   IT systems 34
   TN-C-S (PNB) systems 32
   TT systems 33
restrictive conductive locations 97–8
ring final circuits
   circuit protective conductors 119
   proving continuity 127–8
road lighting columns 139
Road tanker workshop code 144
rod electrodes 6–7
   resistance 10, 12–13
rules of thumb
   CPC current 122
   earth loop fault impedance 65, 66

SELV: *see* separated extra-low voltage systems (SELV)
semi-concealed gas meters 131–2
separate buildings, main equipotential bonding 47–51
separated extra-low voltage systems (SELV)
   automatic disconnection 80
   medical locations 146
separation distance, earth electrodes 15
sheathing, cable
   earth electrodes 10
   earthing and bonding clamps 58
   earthing conductor 21
   main equipotential bonding conductors 40
   supplementary bonding conductors 85
shower cabinets 91
shower rooms 88–91
shows 143–4
small-scale embedded generators 133–6
socket-outlets
   in bedroom with shower cabinet 91
   caravan parks 130–1
   disconnection times 64, 68, 72
   exhibitions, shows and stands 144
   marinas 150–1
   maximum earth fault loop impedance 68
   portable equipment outdoors 75
   significant CPC currents 118
   terminations 113–14, 119–20
   for testing at petrol filling stations 145
soil resistivity 11
source earthing 3–4
special locations 136
   see also locations of increased risk
spurs, busbar systems 120
SSEGs: *see* small-scale embedded generators
stables 94

standby generators, automatic disconnection 79
stands 143–4
static electricity, petrol filling stations 145
static inverters 99
statutory regulations xi
steel building components, connections to 61–2
steel conductors
   armouring, termination 108
   circuit protective conductors 104–6
   galvanic corrosion 14
   k values 156
   main equipotential bonding conductors 42
   rod electrodes 6–7
steel conduit
   earthing conductor 21, 25
   k values 156
steel conduits
   circuit protective conductors 108–11
   main equipotential bonding conductors 40
steel ducting
   circuit protective conductors 111–12
   k values 156
steel grids
   connections to 61–2
   see also metallic grids
steel stanchions
   earth electrodes 9
   extraneous-conductive-parts 56
steel trunking
   circuit protective conductors 111–12
   earthing conductor 25
   k values 156
   main equipotential bonding conductors 40
Stirling engine 136
stock houses 94–7
storage tanks 97
straw storage 94
street furniture 139–41
street lighting 139–41
structural metalwork
   connections to 61–2
   electrodes 9–10
supplementary bonding conductors 84–8
   cross-sectional area 86
   limitations on resistance 87
      agricultural and horticultural premises 97
      bath and shower rooms 90
      swimming pools 92
   supports 87–8, 87–8
   types 85
supplementary equipotential bonding 36, 83–100
   agricultural and horticultural premises 94–7
   bath and shower rooms 88–91
   medical locations 148–9
   other locations of increased risk 99
   portable equipment used outdoors 75
   restrictive conductive locations 97–8
   static inverters 99

   swimming pools 91–4
   where automatic disconnection not achievable 83, 99–100
suppliers' facilities 5–6, 19
supply system earthing: *see* source earthing
supports
   cables in suspended ceilings 142
   main equipotential bonding conductors 43–4
   supplementary bonding conductors 87–8
   see also cable trays
suppressors, protective conductor currents 116
surface area, earth plates 8
suspended ceilings 141–2
swimming pool grids 93–4
swimming pools 91–4
switchgear, metal enclosure or frame as CPC 112–13
system types 27–34
   IT systems 34
   TN-C-S systems 30–1
   TN-C systems 28–9
   TN-S systems 29–30
   TT systems 32–4

tape electrodes 8
telecommunications equipment
   functional earthing 115
   protective conductor currents 116
telephone kiosks 140
temporary supplies 140–1
terminations
   in accessories 113–14, 119–20
   armoured cables 107–8
   earth plates 9
   rod electrodes 7–8
   structural metalwork electrodes 9–10
      testing
   disconnection of the earthing conductor 21–2, 25, 145
   earth electrodes 15–18
   petrol filling stations 145
   protective conductors 153–4
TN-C-S systems 27, 30–2
   caravan parks 130
   earth fault current path 4
   main equipotential bonding 50, 52
   means of earthing 5, 6
   mobile and transportable units 138
   protective multiple earthing 40
   source earthing 3
TN-C systems 27, 28–9
   means of earthing 5–6
   source earthing 3
TN-S systems 5, 27, 29–30
   exhibitions, shows and stands 143
   installations serving more than one building 50, 51
   means of earthing 5
   petrol filling stations 144
   source earthing 3

TN systems 27
  automatic disconnection 63–73
    earth fault loop impedance 63–9
    mixed disconnection times 69–72
      socket-outlets, etc. 72
      using RCD 73
  means of earthing 5
  source earthing 3
touch voltages 35–8
traffic control and surveillance equipment 139
traffic signs 139
transformers
  medical locations 147–8
  see also isolating transformers
transportable units 136–9
TT systems 27, 32–4
  automatic disconnection 73
  caravan parks 130–1
  CPC current exceeding 3.5mA 122
  earth electrodes 6
    limit on resistance 13–14, 18
    loading capacity 15
    testing 17–18
  earth fault current path 4
  installations serving more than one building 50
  means of earthing 5
  medical locations 147
  RCDs in series 76–7
  source earthing 3, 4

underfloor heating 91
uninterruptible power supplies, automatic disconnection 79

vehicles: *see* mobile units
ventilation ducts
  restrictive conductive locations 97
  supplementary equipotential bonding 94
ventricular fibrillation 116–17

warning notices
  exterior gas meters 132
  significant CPC currents 118
  small-scale embedded generators 135
  test sockets at petrol filling stations 145–6
water pipes: *see* pipework
'wet areas'
  circuit protective conductors 101
  see also locations of increased risk
wind power generation 133
wire electrodes 8

Z_e: *see* external earth fault loop impedance
Z_s: *see* earth fault loop impedance